THE CAPER IN SHANGHAI

THE CAPER IN SHANGHAI

LARRY WILES

authorHOUSE®

AuthorHouse™
1663 Liberty Drive
Bloomington, IN 47403
www.authorhouse.com
Phone: 1-800-839-8640

First published by AuthorHouse   08/26/2011

ISBN: 978-1-4567-6638-2 (sc)
ISBN: 978-1-4567-6637-5 (hc)
ISBN: 978-1-4567-6636-8 (ebk)

Library of Congress Control Number: 2011907678

Printed in the United States of America

I get by with a little help from my friends.

The Beatles
In With a Little Help From My Friends

This book is dedicated to my son, Derek Wiles, and my daughter, Misty Lieneginer, but particularly Derek, who pushed me for years to write. He said, "Dad, just write anything, but write." This is also for you Harvey

Thanks, son. This for you.
L.R. (Larry) Wiles

Acknowledgments

I started writing this book in early July of 2009 while on my thirty-second and last official business trip to China. My son, Derek, had been pushing me for several years to write a technical book about supply chain, my field of expertise. A technical book did not interest me; however, the story you are about to read had been running around in my head for some time. If you are thirteen thousand miles from home and can't sleep, why not begin to write that story?

It seems easy—write a novel! Writing anything requires a reasonable memory, diligent research, family, friends, colleagues, and a large dose of help from gracious strangers. Without them, this book would not contain facts where facts were needed and stories to whet my appetite for words to touch paper.

I will be forever grateful to several people who assisted me in making sure my facts about police work, Marine Corps Air Station Beaufort in South Carolina, and air force jargon were correct; reviewing and making excellent suggestions for my factious master clear water filtration system product; teaching me everything I needed to know about ultralights; traipsing around Changzhou and Shanghai China to take pictures of buildings, restaurants, and menus; and translating English to Chinese.

Major Pete Surette Jr. of the Clarendon County sheriff's office in Manning, South Carolina, took time away from chasing bad guys to read the sections of the book pertaining to police work. His guidance was invaluable. Thanks, Pete. Keep chasing the bad guys.

Captain Josiah Nicely, MCAS Beaufort, answered my cold call to public affairs at MCAS Beaufort and was extremely helpful in answering my questions about the air base and making sure I used the correct military terminology. Thank you, Captain, wherever you are in the world today helping keep our country safe.

Brad Noll, Director of Engineering at Wilkins, a Zurn company in Paso Robles, California, helped me design the fictional master clear water filtration system I describe in the book. I almost believed the product did exist as I wrote about it because of Brad's help. Maybe one day it will truly exist.

Jeff Hudson, vice president of engineering at Rotax Engines: I thought about Jack Roush's well-publicized ultralight crash when I decided the same type of crash would play a significant role in my book. Jeff came to my rescue with everything I ever wanted to know about ultralights, including how to crash one. His help made my job easy. Thanks, Jeff. Keep yours flying!

Kevin Ji, director of supply chain at Rexnord Industries, China, a past colleague, was instrumental doing location, building, and restaurant photographs. Kevin also supported my failing memory when I could not remember specifics needed for the book. He also graciously agreed to be my model for Chief Inspector Ji, Changzhou police department. Kevin, I hope my inspector makes you proud.

When I decided the owner of UCPC would speak very little English, like most business owners in China, Bin Wang, senior quality control engineer at Rexnord Industries in Milwaukee, another former Rexnord Industries colleague, graciously agreed to do my translation work. You will find Bin in the book as well. Some of his fictional words are the actual words Bin uses in conversation. Bin, you are a good and patient friend.

To all of you who so graciously allowed me to use your name as a character in the book: I hope you not only enjoy seeing your name in print, but also enjoy the story. You never know; you might turn up in another of my endeavors.

I want to especially thank John and Anne Allen who have significant characters modeled after them. John is my best friend in the world so why not make him a lead character and Anne has been patient with me for over fifteen years as John and I traipsed around the country plying our trade. Thank you both for always being there for me. I truly hope you both enjoy reading about yourselves.

My two children have encouraged me throughout the writing of my book. They had faith in me when many times I did not. Misty and Derek, thanks for pushing Dad on those days when I needed a little push to keep going!

And I also thank everyone else who gave me encouragement to keep going, like Mary Resteghini at the Wyboo Clarendon bank, who asked every time I entered the bank, "How's the book coming?" Dana Montgomery, my doctor's assistant, read some of my initial work and liked it, and Paula Sims, who also works for my doctor, checked each visit on my progress. Rob Johnston kept the e-mails flowing, asking when he could read the book.

Finally, to you, the person who purchased this book: It is my fervent hope you will enjoy it and continue to read my coming work and those of other authors. Reading is the key to knowledge. Turn the lock and come on in. My second book is on the way, and a third is cruising around in my head.

L. R. (Larry) Wiles

Cover photograph by Sarah Less of Back Jacket Photography

Chapter 1

How could the sound of a ringing telephone cause a quiet, early summer Southern morning to explode like a hurricane, catapulting me into intrigue far from home and almost getting me killed, all for just trying to help my best friend! When you are comfortably retired, the thought of spending weeks in Shanghai for anything other than exquisite, diverse cuisine plus ancient Chinese culture is not really appealing, but everything that follows did happen! Join me on my journey, listen to my story, but stay close, because I am only going to tell it once.

My name is Richard Randolph Watson III. My parents hung that one on me when I was born, and thank goodness it never stuck! My friends call me Rick.

It was unusually hot for the month of May on Lake Marion, slightly north of Charleston, South Carolina, that fateful morning. When I say hot, I mean hot! There is nothing like a brutally humid South Carolina day to make you slow down, drink loads of sweet iced tea, and fan. Something all Carolinians are born to do!

My wife, Terri, was sitting on our spacious screened porch with four ceiling fans moving volumes of hot, moist air. It was mid-morning on Tuesday, May 6, 2008. Vito, our well-fed black and white cat, lay on her lap, purring while dreaming about whatever cats dream about when they sleep. Two years ago, Vito, exhausted and dangerously under nourished, appeared at our back porch screen door, meowing weakly and looking as though this was his last chance for survival. We did our best to nurse him back to health, and it worked. He adopted us and instantly made himself at home. It quickly became obvious Vito thought all humans were just warm furniture and the perfect place to lie.

1

Our porch overlooked the lake where our twenty-three-foot Harris pontoon boat and Honda and Sea-Doo jet skis waited patiently at the end of the dock. The jet skis sat waiting for someone to crank them to life and explode across the lake, making forty-mile-an-hour banks left and right, cascading water across the lake like an oil gusher reaching skyward after a big strike. The pontoon boat was for fishing and a more leisurely lifestyle. The fishing members of the family had managed to relieve the lake of several large Arkansas blue catfish on late night expeditions that were also for sharing bodacious stories of fish long ago caught or miraculously lost after exhausting fights for survival.

I had just returned from brim fishing at the end of our cove in our johnboat I called the *Brim Buster*. The johnboat was my favorite. It had a flat bottom with a wide beam to accommodate shallow water that filled the slews and coves around the lake. The previous owner had added padded seats for long hours of patient waiting and camouflage paint for duck hunting during the fall and early winter. I planned to repaint the boat a more respectable fishing boat color, but the camo had grown on me, and besides, I hated to paint! Every spring, brim returned to shallow water to construct beds for spawning a new generation to entice fishers who fished the lake. Brim beds also were built under the docks in the cove, but were abandoned early in the spring. The best fishing was always at the end of the cove. The big ones resided there . . . The *Brim Buster* was equipped with a Mercury fifteen-horse outboard and a small trolling motor. The trolling motor was critical for maneuvering deep into heavy grass at the end of the cove where the brim beds were and larger brim swam, just waiting for me.

Annabell, our nine-pound red Dachshund, was my faithful fishing companion and good luck charm. She was constantly in motion in the boat, looking ahead and behind. Scanning the heavy grass and flowing water like an electronic fish finder, barking when she determined the right spot for me to drop my line. Annabell always picked just the right spot! Her intuition was uncanny, and I learned to trust her completely. She knew it was fishing time when I snapped her into her lifejacket. Her confident little butt and tail wagged with joy as we headed toward the boat.

Today had been a good day. We managed to pull thirteen palm-sized brim into the boat. The best size for frying in Terri's secret special corn meal recipe. It would be good eating tonight with Southern potato salad, coleslaw, hush puppies, and Corona with lime.

Terri was drinking coffee as I watched. Now, I know there is folklore regarding drinking coffee to cool when it is hot, but it never worked for me. Iced tea, good old-fashioned Southern sweet iced tea, was what I craved. The so-called experts who created the folklore had never experienced a South Carolina summer!

I stood there, out of her sight, watching and admiring her. She was quickly approaching her fifty-eighth birthday, and if I do say so myself, she put most women her age to shame. She could stroll any beach and still turn heads. Her short golden hair was the color of wheat ready to be harvested, and her smile melted summer ice.

You would never know that she had raised two girls to adulthood and had three grandchildren.

Life had not been easy for her in the early years. Her first marriage in 1970 at nineteen ended in divorce in 1973when her husband quietly left town with his girlfriend, leaving her nothing and abandoning his daughter, Judy, both emotionally and financially. A second marriage in 1974 ended in tragedy in 1978 when her second husband was killed in a car accident on one of Charleston's narrow, moss-covered, two-lane roads that had claimed many other lives in similar accidents. It was a head-on collision with someone driving too fast after a night of heavy drinking at a local bar. Mary, her second daughter, was two, and there was no insurance to cover the costs of the funeral.

Moving from place to place around Charleston with her girls had been a ritual after the accident because of low-paying jobs that created a lack of funds to consistently pay rent. There had been many days where it was not a sure thing there would be food on the table or new clothes for the girls, let alone milk.

It was a month-to-month struggle. They spent an occasional month in apartments within close proximity of the beach on the Isle of Palms or Sullivan's Island, Terri waiting tables in restaurants when jobs were available, and the tips were plentiful; however, most months were in apartments that you stayed inside for safety and prayed for the next job to pay more than barely a daily subsistence.

Her first stroke of good luck occurred in 1979when her best friend convinced a station manager at the Avis Charleston airport rental car location to employ her to shuttle rent-ready cars from the Avis off-airport lot to airport customer pick-up locations. She quickly progressed to counter leasing agent, making more money than she dreamed possible.

A promotion to Assistant station manager in 1982 and a string of awards recognizing outstanding customer service finally provided stability to her and the girls.

The girls, now thirty-seven and thirty-two, trusted their mom completely growing up and never really knew how bad it was at times. Peanut butter was a staple, and to this day, neither daughter was a big fan.

Judy had gotten married at sixteen to her childhood sweetheart and lived in a suburb of Charleston. They had two children: Bill, now twenty; and a daughter, Jill, now eighteen.

The younger daughter, Mary, now thirty-two, was married for the second time and the mother of one son, Jamie.

Terri and I met in July of 1994 when I rented a car on a business trip to Charleston, and she was working the counter, replacing one of her agents that had skipped work that day. Falling in love with her was easy, and I made frequent trips throughout August just to rent cars as I summoned the courage to suggest dinner in Charleston. She reluctantly agreed, but made sure I knew this was not normal for her to date customers—not that she hadn't had ample opportunities. Charleston airport was frequented by celebrities who pursued her while visiting the city or the beaches and golf courses that were notably famous throughout the country. Hot-shot business guys also fell for her beauty and Southern charm. One afternoon in 1983, she rented a car to Danny Ford, a legendary Clemson football coach, who flew into the Charleston airport from a recruiting trip. He was scheduled to speak that evening to the Clemson Touchdown Club of Charleston. Coach Ford invited Terri to attend the event. She accepted and from that evening forth became a passionate follower of Clemson football. Her meeting with Coach Ford also rekindled a long-lost desire to complete her college degree. Terri entered Clemson in the fall of 1984, and six years later, after hard work at nights and on weekends, graduated from Clemson with honors with a BS in business.

The night of our first date, I arrived at the home she shared with one of her younger sisters, her mother, and daughter, Mary, promptly at six in order to arrive at Poogan's Porch in downtown Charleston for a seven o'clock reservation. I had strategically selected Poogan's to impress her with my knowledge of high-quality restaurants in the city that matched her cuisine favorites.

I melted into a puddle of freshly churned butter when Terri walked down the hall from her bedroom to the small but very comfortable living room. She was the vision I had searched for to share my life. She was wearing a black and white pantsuit that was fashionably correct for the times and styles. There was no doubt in my mind that we would never again be apart.

Conversation flowed like a cool mountain stream as we drove toward the city on Interstate 26 during a Lowcountry summer thunderstorm that instantly appeared from a deep, dark black cloud, that produced sky-to-turf lighting and deposited copious amounts of rainfall and quickly departed the scene. We laughed when we both spotted a small fishing boat on a trailer with a flat tire stuck on the opposite side of the expressway containing one lone passenger sitting in the boat holding a small umbrella over his head in a futile attempt to keep dry.

Dinner exceeded my expectations and impressed Terri as my suggestions of dishes were her favorites. Our compatibility was instantly obvious as we discussed a wide range of topics from our individual businesses, our thoughts on the current state of political affairs both local and national, our mutual love of sports, and raising children.

As I predicted, we were never apart from that first date in 1994 and have been happily married for fourteen years.

Both of us had decided to retire in September 2007, nine months ago. Terri had been responsible for inside sales and customer service working for Lincoln Lift Truck and I was vice president of supply chain for Wright Manufacturing. I accepted a position with Wright after retiring from the Marine Corps and had worked to the Vice President level. We were financially comfortable and no longer had anything to prove regarding our reputations being successful businesspeople.

I graduated from Murray State, a small university in Western Kentucky, in 1965, my schooling paid for with an ROTC scholarship. I was a catcher on the baseball team. I loved baseball from early childhood, and by the time I was eight, I was considered a potential prospect.

My father was employed by the Illinois Central Railroad when we moved to Centralia, Illinois, perfect small-town USA, when I was eight. My childhood was spent on the baseball field. Baseball began early each hot summer morning and ended with the all-important little league game in the late afternoon. Mom made sure my uniform was sparkling clean or at least as clean as possible for each game. I became a major leaguer wearing that uniform. I returned it to her after each game, win or lose, pants with hard-earned dirt streaks down my sliding leg from dashes to second base

5

trying to break up a double play or grass stains from making what in my mind were spectacular diving catches on sacrifice bunt attempts in front of home plate. My top with "Moose Lodge" lettered across the front would be filthy from sliding on my belly at home plate to score a run and sweat stained from wearing the all-important protective catching gear.

My friends and I breathed baseball twenty-four hours a day, fantasizing how our heroes would have done it and doing our best to emulate them. Del Rice, a journeyman catcher who played for the St. Louis Cardinals, was my hero. I met him on my eleventh birthday when my father took me to see a Cardinals game as a birthday present.

I pushed my way to the front of the Cardinals dugout before the game began and begged him for his autograph. He took my Del Rice baseball card from my outstretched hand. It shook so hard I nearly dropped it. My voice creaked like an adolescent voice that is changing to manhood, but I managed to squeak, "Mr. Rice, I'm a catcher just like you."

A smile erupted from his weather-beaten face, and he replied, "Keep it up, kid, and you might be signing baseball cards one day." That baseball card became my prized possession and inspiration for each game I played. I still have that card today.

A few major league baseball scouts approached me during my senior year in college, but an ROTC commitment required I report immediately upon graduation to officer candidate school in Quantico, Virginia, for basic training and subsequent deployment into the Marine Corps. I hoped my military occupational specialty would be part of a military police unit, and I was lucky. I spent my career in military investigations, rising to lt. colonel prior to my military retirement.

Retirement from industry had been difficult at first. It was difficult to remove ourselves from the everyday speed of business to the tranquility of the lake, but the golf courses, the pool, and the lake toys had taken hold.

We also spent as much time as possible with my daughter, Misty; her husband, Rob; their three children; my son, Derek; and his wife, Geraldine. Misty lived in the steel city of Pittsburgh, and Derek had moved from San Diego to Boston after a successful career in the technical side of theater to open his own well-regarded special events company.

Terri and I were always physically active and in a physical condition that belied our ages of fifty-eight and sixty-eight.

Terri was a record holder in multiple sports in her Harleyville, South Carolina, high school, including track. She won titles in several running events as a junior and senior at the high school state track meet in Columbia, the state capital. Terri returned to running later in life and became a fervent tri-athlete recognized across the South as a winner and leader, winning in record times and representing tri-athletes as president of the Tri-Athletes Association of South Carolina.

I was also a runner. I had always been a runner. A day without running created a sense of anxiety within me. Running also produced the perfect opportunity to blend conditioning and problem solving. Some of my best problem solving was generated during my long slow runs. I had never been a speed merchant, preferring to let the environment absorb me while I thought.

We frequently competed in marathons throughout the South, always placing in good positions for our individual age groups. Our daily early morning runs across strategically selected lakeside routes allowed us to immerse ourselves in the magic of the Lowcountry. Mornings where fog caressed us, a mother deer protecting her newborn fawn, startled by our footsteps, showed us her tail as she and her fawn bounded across an open field; hot mornings smelled of lakeside plough mud and decaying vegetation.

A ringing phone jolted me back to reality and I hustled into our great room to answer it.

The call lasted only a few minutes and left me in a complete state of shock!

Terri looked up as I returned to the screened-in porch. "My God, Rick, you look like you have just lost your best friend."

"I may have," I replied. "That was John Alworth. He is in China again and says he desperately needs our help."

John was better known as "Little John" and had been my golf foil for years. He had been our friend since the days John and I ran plants next door to each other in Charleston.

John was now the director of quality for Water Management Technologies, a fast-growing, privately-held company that had experienced over 500 percent growth in the past two years and was on the verge of going public. Their new innovative product was designed to control all the water applications in a consumer's home with leading-edge green technology, and when it came to market, it would be cited as a model in the global warming frenzy.

WMT had also been the pioneer in using Chinese suppliers to produce their current systems. John had been there five years ago when the new product was first discussed and was appointed to the position of quality leader for the initial design and system production. He was responsible for assuring the new product met the rigid Water Management Technologies quality and engineering specifications required when WMT produced the systems the American public would be using in their homes. The company's strategy was to infuse their designs into office buildings, hotels, and large sporting venues as soon as the market for consumers had been established.

"What has he gotten himself into this time, and what is he doing in China?" Terri asked. "I thought he was through with traveling there and moving to Georgia."

I replied in a stunned tone, "I don't think his mind is on moving to Georgia today; it seems that he has been charged with murder!"

"What! You're kidding; how could that be?" Terri replied. "He wouldn't hurt a flea, let alone another person!"

"There must be some mistake."

"Wait a minute: are you sure he is not just trying to get your focus away from the Clemson-Carolina charity golf tournament? You know you can't trust him when the tournament is involved."

The Clemson-Carolina charity golf tournament was the second most competitive event held between the universities with the yearly football grudge game, traditionally played on a Saturday afternoon in late fall, easily number one! Clemson and South Carolina, located less than two hundred miles apart, were bitter rivals to say the least, and the Clemson-Carolina football game was the ultimate in school rivalry. Pure hate permeated the landscape the week prior to the game!

Each football season, every player and coach fervently swore the current week's scheduled contest was the most important game of the year, but everybody knew it was bunk! The outcome of any event, athletic or academic, must embarrass the hated Tigers or Gamecocks.

The football game was always the last game on either school's schedule. Records meant nothing; just win and the season was considered a success no matter the team's final won/lost record. Winner's bragging rights would reverberate through split families, in the workplace, and across the other competitions between the two rivals.

Alumni, students, and just fans who identified with either Clemson or Carolina meticulously planned this weekend. Tailgate parties, backyard barbecues displaying garnet or orange flags flying gallantly high above were symbols of the faithful.

Bright orange sweatshirts, orange hats, and women's dresses in all orange reflected off sunglasses in the mid-afternoon sun. Fans in garnet-colored clothing with the Gamecock logo and "cock" banners dominated the other half of the landscape.

You see, my wife was a Clemson graduate, and John, a graduate of the University of South Carolina. Both were very proud of their alma maters. Each football game was just another opportunity for Terri or John to plan a diabolical practical joke on one another! As each Clemson Carolina football game approached, the taunting and scheming between them increased. One particular pre-game practical joke still remains at the top of the extensive memory list.

My dear wife only drove a car. She had no earthly recognition that cars needed anything but fuel to support her daily activities. To lift the hood would be a miracle of magnificent proportions.

John had secretly attached a Gamecock license plate on the front of Terri's Buick Regal, and she meandered around Charleston for weeks showing her unknown support for the hated Gamecocks. Reality struck hard as she dropped her purse while passing the infrequently visited front of the vehicle, and to her amazement, she realized that her car with orange paws placed strategically to draw the attention of even a casual observer to her passion for Clemson had been desecrated with a front Gamecock license plate!

"How could he!" she related later. Paw Power raced through her veins as she spun in place and stormed like a boiling tropical storm searching for her cell phone. "How could he!" She spat over and over as she punched each number on the cell phone keypad with vengeance.

Coincidently, Little John had just arrived in my office for a meeting with me to discuss a situation that affected both of our plants. A tidal creek slithered like a snake behind both plants and frequently became the home of baby alligators pushed from the safety of their mothers by the flowing tide. A ringing telephone broke our conversation.

I punched my speaker when the telephone rang, apologizing to John for the interruption. A voice exploded as the speaker vibrated across the desk. "Is that little asshole with you?" Terri screamed.

Laughing hysterically, John realized Terri had found the Gamecock license plate. John's retort was simple and extremely effective: "*Go Gamecocks!*"

The charity golf tournament was no different from any other Clemson-Carolina contest! The tournament always contributed hefty amounts of money to the chosen charities of each school with a sizeable and increasing turnout each year. Even a casual observer knew which course was hosting the tournament with the profusion of orange and garnet dominating the golf course and each entrant's colorful golf equipment!

Terri was active in the planning and managing of the tournament from the Clemson side. John and I had taken turns winning the tournament over the last several years with foursomes of equally skilled golfers always up for the challenge! The coming tournament would not be an exception. John's team had won last year's tournament, besting my team by one stroke. John rolled in a thirty-foot putt through a double break green on the scintillating eighteenth hole to win the match. I, like the football coaches, worked a full year plotting my revenge by urging my foursome to work relentlessly on their golf game and not forget last year's outcome! I was confident my Clemson team would be victorious this year!

The golf tournament was also special for John because he met his current wife, Ann, during the tournament three years ago when Ann acted as a player hostess as part of the Carolina contingent. John had been single since his first wife, Betty, had died a few years ago of lung cancer. John was devastated. Terri and I had been his touchstone during this dark period in his life. We worked to assure we were always there when he needed us. Never pushing, not suggesting, just there when he needed to talk or vent his frustration or share his overwhelming grief. Betty and John truly loved each other, and it took several years of introspection for John to recover.

Ann was assigned to John's foursome plus one other that day, but the other foursome never really saw her after the first hole. To say that John and Ann swept each other off their respective feet was an understatement! Ann at forty-two was twenty years younger than John, but that never mattered.

John was not just smitten with Ann's dark Italian complexion, profound flashing brown eyes, and cascading auburn hair that fell like a waterfall from her shoulders reaching to her mid-back. He was instantly in love! To put it mildly, Ann was always the sexiest woman in the room. Ann could hardly control her own golf cart for looking at John and absorbing his every move into the deepest caverns of her brain. By the eighth hole, Ann was ready to find a pastor and make her feeling official! A whirlwind romance followed, and they were married on the Cayman Islands on January 9, 2005. Terri and I gladly accepted the responsibility as best man and bridesmaid.

Initially, the four of us were inseparable, but John's new position at WMT and his consistent travel to China had caused less frequent dinners, golf games, shopping trips, and vacations between us.

Ann frequently accompanied John on his trips to China and had developed many new friends in Shanghai. She had started a small business acquainting wives of US professionals newly assigned to positions in Shanghai with the city and Chinese customs.

Terri and I loved John, and over time, we felt the same way about Ann; however, Terri worried that Ann was, at times, exceptionally vain about her appearance and always conscious of what others thought about her and John's lifestyle. Everything had to be top of the line. The biggest house, the most expensive cars, a large boat that hardly left the dock, and expensive trips were the norm until John's involvement began in China.

Ann was always pushing John to earn more to pay for her ever-increasing lifestyle. I was also somewhat concerned about the truthfulness of a variety of Ann's comments but wrote off my concern as just a reaction to Ann's driven enthusiasm.

Ann enjoyed her new business in China. She had definitely become somebody to her corporate clients. Many of the newly arrived wives believed only Ann had the vital contacts to quickly solve any personal problem that occurred. They trusted her completely.

"No, I don't think this time John was screwing with me regarding the upcoming tournament," I said, replying to Terri's question regarding John possibly trying to disrupt my focus on the golf tournament. "He sounded desperate and scared as hell on the phone."

"Where was Ann? Did she go with him on this trip?"

"Oh yeah, she was in the background sobbing and yelling at me to do something!"

Terri looked at me with a quizzical look on her face. "What does he think we can do about it? We aren't associated with any police or government law organizations, and your time as a military police inspector in the marines isn't the type of credential that will be welcomed in China. What is his company doing about it, and what about the US embassy? Surly they can do something to help."

"He told me that he did not know where to turn—that his company may be involved in something that if found out would destroy not only the company but cause the loss of the lives of their customers. Zhu Zhong Huang, chief engineer for Universal China Production Company, the main supplier in China located in Changzhou, met with John at the factory when he arrived there yesterday. He told John there were fatal flaws in the design of the new product but would not go into any details while in his office. He was to meet John at his hotel in Shanghai later that night but never showed. John said he was worried when the guy didn't meet him, but he was suffering jet lag and fell asleep in his hotel room."

I continued, "Later that evening, the police arrived at John's hotel room banging on his door. Still half asleep, he answered. The police entered his room and not so politely informed him that Huang was dead and that they had evidence that John was the killer. Before John could respond, he was escorted to a holding cell in a jail in the center of Shanghai. As of now, they have presented neither details nor evidence implicating John, but they did allow him to make one telephone call, and he called us.

"I asked John where Ann was while this was going on, and he told me she had arrived a few days ahead of him to meet with one of her new clients. Evidently, she was having dinner with the client when this happened, but she did make it to the police headquarters. Ann suggested John call us."

"Oh dear, "Terri replied to my description of the call. "Does that mean that we are on our way to China?"

"Guess so. Not sure what we can do, but you know we can't let him down."

"By the way," Terri asked, "if the police will not provide any evidence that John is the killer, why do they think he is responsible, and why didn't he call his company lawyer?"

"All he said was that he didn't do it and muttered something about Gregory Brightson, the CEO. John said, 'Gregory would do anything to protect his company during the planning to go public timeframe, and Brightson on cannot be trusted.' I was his best friend and the only person he could trust to help him prove his innocence!"

Gregory Brightson, better known as "Chalk," a name given to him as a young schoolboy, was the charismatic leader of Water Management Technologies.

Gregory was born in the small midwestern town of Kingstree, Indiana. Kingstree was a small farming community with one large automotive parts plant as the main town support. He had been a star at an early age. Gregory had walked early and talked early, and his straw-blond hair and sapphire-blue eyes caused all who interacted with him to predict big happenings in his future. As he grew older, this fact was not lost on him. Gregory truly believed he was one of those very special people who were on a mission to greatness.

In grade school, he was fascinated with Mrs. Jenkins, his third-grade teacher and her control of power using chalk on the chalkboard. Young Gregory was always the first to volunteer to write his answers on the board and use the powerful chalk. He was the best chalkboard cleaner ever and carried his love of using a chalkboard throughout his life. All conference rooms at WMT, as well as his office, contained chalkboards and stacks of colorful chalk.

His best friend in the third grade, Mike Love, was the first person to dub him with the nickname "Chalk." Gregory liked the name and insisted that be his name from that day forward. The name stuck and followed him through high school and college, and then into the business world. His good looks and outrageous blue eyes made him a favorite with women of all ages. He developed a scandalous reputation during his college years and during his business life. He had no time for commitment but enjoyed the company of all types of the fair sex. Several women tried to claim him, but each one failed.

Gregory was always the star. In class, he always attained the highest grades. On the athletic field or floor, he was the fastest and the most accurate in any sport he tried. He was captain of the football team at Southern Indiana, a small liberal arts school. Gregory was offered full scholarships to larger, more prestigious universities, but his father was a graduate of Southern Indiana and pushed for Gregory to follow in his footsteps. The St. Louis Cardinals drafted him as a hard-hitting outfielder with a major league arm, but Gregory decided during his senior year his mission was in industry and the development of a company that would make him wealthy and immortal. To this day, his mission was still intact.

He started Water Management Technologies on money borrowed from his father and his small town bank. When he needed more money to propel the business forward, he met with several venture capital companies. The old saying "sell ice cream to Eskimos" was true when it came to Gregory.

His concept of managing energy around water in homes and public buildings was a radical change and was embraced by several VCs. His dynamic personality just made potential investors feel they could not lose with the young lion standing in front of them laying out his strategy and their financial opportunity.

Employees at WMT were also fiercely loyal to Gregory. They would follow his lead without question. Those who were employed early, like Little John, knew they were also on the fast track to riches. Newer employees worked hard to catch up, knowing they also had a bright future ahead. Water Management Technologies was on the way, and every employee wanted to be there when the big day arrived! Chalk would lead them there!

Terri said, "You book the tickets, and I will call Mary to watch Annabell and Vito. How long should we pack for?"

"Not sure," I replied, "but at least three weeks. Hell, we may get there and find there is nothing we can do but add moral support, but we have to try. That's the least we can do!"

Chapter 2

There aren't many choices from the Charleston International Airport to anywhere in China. You flew Delta. It took us two days to find available seats in business class costing thirteen thousand dollars each from Charleston on Delta flight 2700 to Detroit, connecting with Delta flight 25 to Shanghai, leaving Charleston International at 7:00 a.m. on Friday, May 9, and arriving in Shanghai at 10:30 p.m. on Saturday, May 10. Fourteen hours nonstop from Detroit: the time was filled with movies and food and whatever sleep was possible. At least the seats reclined flat, somewhat resembling a bed.

Terri slept while I watched movies and thought about John, trying to determine where to start upon our arrival in Shanghai, how to find a way to prove John's innocence and return our lives to normal. I knew I also needed someone from Shanghai to support me.

I telephoned Bin Wang, an old friend in Shanghai, prior to our departure.

Bin, like most Chinese business professionals, had adopted an American name to make communication with each other easier for Americans who always have difficulties pronouncing Chinese names. Bin, forty-one, was the first person I hired when I opened Wright Manufacturing's first office in China. He had assisted me in recruiting the professional talent we required and ran the office as director of supply chain in China before leaving when I retired to start his own company, Wang and Associates, finding quality suppliers in Asia for American companies coming to China to purchase products and assuring development of business relationships.

I was confident he was just the right person for the job. He was afraid of nothing and had been my right hand on several major negotiations with suppliers in China over the past several years. Bin was an outstanding negotiator of contracts to purchase products. We had successfully increased Wright's purchases in China over the years in excess of $300 million and generated cost savings of over $45million.

As the night progressed and sleep was distant, I remembered an event in my marine investigative past in 1998 I hoped could be used as experience to help me solve John's case and prove his innocence. I had been involved in a major drug case working undercover at Marine Corps Air Station Beaufort in South Carolina.

I couldn't sleep, so my mind drifted back to that case.

The air station, better known as MAG-31, was located in the heart of the South Carolina Lowcountry. I had volunteered to work undercover there to solve a drug case that was affecting the base.

MAG-31, currently home to six Marine Corps F/A18 fighter squadrons, was reactivated and moved to Beaufort in 1961 after being decommissioned from Marine Corps Air Station Miami in1958. MCAS Beaufort's flight line with its concentration of marine fighter aircraft came to be known affectionately as "Fightertown" to the pilots and crews who worked there. Many of the marines' best pilots trained at MAG-31 at some time in their careers, and some of them returned as instructors to ensure that the best became the "best of the best." Many of them fell in love with Beaufort and the surrounding Lowcountry, including Charleston, returning to retire and to run local area businesses. Though not a pilot, I was one of those who chose to make the Lowcountry my home.

The base was located on 6,900 acres on what were old historic plantations in the South Carolina Lowcountry. The area was rich in early American history. In 1779 during the American Revolutionary War, Union troops landed and occupied the area around Laurel Bay, which later became the future site for military housing for families assigned to the base. A major battle was fought at Grays Hill, also on the base. The base had grown over the years and now housed 4,000 marines and sailors and 700 civilian employees.

I arrived in Beaufort on Friday, January 2, 1998, fresh from a posting in Cherry Point, North Carolina, at 2nd Marine Aircraft Wing headquarters. It was a cold, wet winter morning, normal for Beaufort that time of year. This was my second posting at MAG-31, albeit scheduled to be temporary this time.

I was assigned to Colonel Jack Jackson, the base commander. Colonel Jackson was a marine's marine. He was six feet, three inches with his salt and pepper haircut into a military flattop and dominated any room he entered. Jack and I had worked together before in Korea, and he had specifically requested me as part of his team to assist him in breaking up a major drug ring that involved marines stationed at MAG-31.

My job, as I understood it and based on the data the colonel had provided me to read on my flight in, was to go undercover with fellow technical training officer Major Gary Hildegard. Major Hildegard was scheduled to arrive at MAG-31 also on January 2. So we would not be known to anyone on the base. I was to operate, as the new owner, a rather rundown biker-type bar with a reputation. The bar had recently been secretly purchased by the Marine Corps as a cover for our operation. I was to hire Major Hildebrand as a bartender upon his arrival in Beaufort once I was in place and operating the bar.

This was something new for me. I had investigated everything from thefts of top secret documents, missing persons, thefts of military equipment, and even deaths, but I had never been asked to work undercover. I was apprehensive but excited about the opportunity to add this type of assignment to my resume.

"Lt. Colonel Watson is in the outer office to see you, sir," the lieutenant on duty said as he interrupted Colonel Jackson's thoughts.

"Send him in, Lieutenant," Jackson replied, being brought back to the day's reality.

"Lt. Colonel Watson reporting for duty, sir!"

I was right behind the lieutenant as he entered the room.

"Holy shit, Rick, you look great, and it is wonderful to see you!" Colonel Jackson shouted. "How's Terri?" Not giving me time to answer, he continued, "I am sure glad you could make it," giving me a bear hug with such force it was apparent age had not diminished his physical strength.

Colonel Jackson crossed the room and flopped down in his high-backed leather swivel chair behind his desk after another round of handshakes and questions about my flight in and the adequacy of my room in the visiting officers' quarters. "I sure need your help on this one!"

The colonel's office faced marsh land that is famous in the Lowcountry and was very well appointed with a large mahogany desk with appropriate side chairs, conference table, two couches, and bookshelves containing reference material and mementos of his military and family life. I took a minute to review the various decorations and to relive memories of my relationship with Colonel Jackson.

I slid into one of the leather chairs across the desk and began the conversation. "I read the data you sent me on the plane. It sure looks like this is one big case and needs to be cleaned up before the press gets wind of it."

"You know, Rick, I have been the base commander here for three years, and this base has been a model for the Marine Corps. We take the best fighter pilots from throughout the corps, the cream of the crop, and improve their skills as pilots to become the best air jockeys in the world, and then send them out to the Atlantic or nowadays straight to the Middle East. Never a problem! Top quality fighter pilots all with some damn fine crews keeping them that way!

"All of a sudden things have changed," Colonel Jackson lamented. "The quality of the fighter pilots has improved, but the enlisted marine is a different animal. Our average marine is less educated, feels he is entitled, and really does not want to work hard. Some of them believe after Parris Island, the worst is over, and the rest of the enlistment is a way to make a paycheck!"

The colonel continued, "Of course, I am exaggerating and clearly personally disturbed over this current situation, but we need to continually work to weed out the bad marines and encourage the good marines to take the lead. Our mission here is critical for the country's survival. Our guys are always the first on the scene anywhere in the world, and we depend on them to control any situation we find ourselves in. We must continue to be the best to make that happen."

"We can't afford drugs on this base. Can you imagine what could happen on our flight line if some of these guys were under the influence! We have random screening, but it is a constant battle to keep ahead of the technology for circumventing it. Oh, in the past, we've had the occasional problem with a marine getting involved with the wrong people and getting hooked on cocaine. Weed is always there, but peers mostly handle that issue. Thank goodness we haven't seen any occurrences of meth. I believe this current situation is becoming an epidemic and has happened so quickly that we didn't see it coming. I'm certain most of my officers don't realize the severity of the problem."

I thought for a moment and asked, "Have you made any arrests related to the latest incidents?"

"Of course, we have made a few arrests," Colonel Jackson replied, "but they are just the small guys, the type that always get caught! The escalation of arrests is what finally tipped us off to the problem. I pushed Cherry Point to elevate this one in order to get to those behind the operation. I selected the bar and requested you to help me. Major Hildegard was recommended by my old friend Colonel Billy Williams, and I trust him explicitly. I believe Major Hildegard will be the right partner for you to break this case. I have asked him to join us at the officers' club later tonight so you two can meet."

"Have you developed any possible informants we could work with?" I asked.

"Informants have been hard to come by. Whoever is behind this operation seems to have the power to keep potentials in line. We do have one corporal who was arrested last week that may be willing to cooperate for the right deal. I have requested approval to commute his sentence if he cooperates. I should have approval early next week."

"What about the bar? What happened to the current owner, and how will we explain the change of ownership from him to me? Won't the locals be suspicious when he leaves abruptly and I appear?" I asked the colonel, trying to get my mind around what he was asking of me.

The colonel replied, "We were lucky on that one. We identified the bar as a possible distribution point for drug activity. It's a hangout for our marines after hours, and some pretty low-life characters frequent the place. There is one person in particular who is of interest, but nothing solid. Only by rumor and reputation. The current owner, Jim Baldwin, is in deep shit with the IRS for unreported income and back taxes, so we made him a deal he couldn't refuse, as they say. He agreed to work with us understanding if he weasels, it's jail and a lost key!"

The colonel continued, "You will be introduced as his cousin from Missouri, and Major Hildegard will be a guy looking for a job as soon as you take over. The main thing is to paint the picture that you are just enough from the wrong side of the tracks to establish credibility. Phil Knollson, the local chief of police, will make a not so friendly, timely visit as soon as you are in charge to help set the stage.

"Rick, I have a meeting with the Beaufort mayor on an unrelated matter in a few minutes. Let's continue this discussion tonight when you meet Gary at the officers' club, if you're in. I know this could be a very dangerous assignment, and I would fully understand if you said no, but I sure do need you."

"I'm in, Jack. I'm a big boy, and I have always been able to take care of myself, but I appreciate your honesty and concern," I replied. I looked at my watch. "It's sixteen hundred now; how about I get my quarters organized, call Terri, and meet you about eighteen hundred at the club. Okay?"

"Okay, I'll let Gary know, and thanks, Rick; I knew I could count on you."

I left the colonel's office and headed back to my quarters. It didn't take long to unload my bag and organize it. I called Terri to catch up on her daily events, and in an hour with everything complete, I decided to spend the time before our eighteen hundred meeting reliving old memories of my last posting at MAG-31.

My route took me past the newly constructed enlisted barracks that would house 238 more marines. I learned later that there was another barracks scheduled next year that would house an additional 211 more.

I passed one of the several softball/baseball fields strategically placed on the base and spent a few minutes watching a hotly contested softball game, remembering my own time on that diamond in just the same type of game.

The fitness center was as active as I remembered it, as marines of all ranks worked to keep their fitness at a top level.

My final stop was the fishing pier where I spent many an early morning and evening attempting to lure grouper and bass onto my hook. Sometimes, success meant a well-prepared dinner was coming.

During shrimp-catching season, my partner and I would spend fruitful hours casting along the banks of the tributaries for those large, pink South Carolina shrimp that would cause neighbors to hang over the fence as the shrimp were barbequed on skewers basted with one of my very secret seasonings!

It was almost eighteen hundred hours before I knew it, and I had to hustle to make it to the officers' club on time.

It was 1802 when I entered the club, and the "Hornets Nest" was just as I had remembered it. A fifteen-foot dark mahogany mirrored bar was showcased on most of one wall of the club. The comfortable, low-lit seating and dining area of the club also contained dark mahogany tables and chairs, and was dominated with leather easy chairs and couches. I always thought some Washington procurement agent negotiated a very large cost reduction to place this type of decor in many of the marine officers clubs I had frequented.

The walls were decorated with a eclectic mix of Lowcountry pictures and paintings, and air shots of the various planes that have been based at MAG-31. One wall was filled with an expansive mural depicting the history of both the Beaufort area and the growth of the base.

I spotted Colonel Jackson and Gary Hildegard sitting at the bar, both drinking a beer. The colonel swung around on his stool as he felt me approach. "Glad you could make it, Rick," he said. He turned to the person sitting beside him. "This is Gary Hildebrand."

It only took a few seconds to assess the major's physical presence. He was dressed in civilian clothes, short, maybe five six, balding with glasses, but had the physique of a small college linebacker and the look of someone you always wanted on your side.

"Nice to meet you, Gary," I said. I shook his hand with my firmest handshake.

"You, too, Rick," Gary replied. He increased the pressure on my hand. "Colonel was just telling me about how you two met and a few of your adventures."

I gave Colonel Jackson the eye and replied, "Don't believe most of what he says unless it's about me. He always exaggerates his portion of those adventures, as he so fondly recalls them!"

Colonel Jackson laughed, returning my look. He said, "Let's move to a table in the corner that is more private so we can discuss the case and the operation."

I scanned the room and located a four-person table on the north wall next to an emergency exit that was unoccupied and fit the description. "A corner table it is. How about that one?" I said, pointing toward the north wall. We finished our beers and headed to the table.

Three briefing books were strategically placed on the table when we arrived.

"How did you . . . ?" I exclaimed, looking at Colonel Jackson and the briefing books lying on the table.

Colonel Jackson just smiled and said, "Let's get to work."

The colonel began, "What you have before you is a briefing book with everything we know about this case. It has been compiled in cooperation with the local police, the DEA, and my base MP investigative unit, and it is considered to be top secret. We have tried to include every detail we believe you both will need to work this case. If we missed something you think we need, just tell us, and we will get it!"

He continued, "The book in front of you contains five sections with the following data:

"Section one: A compilation of the suspects who frequent the bar. It contains biographies, pictures, criminal records. They are organized from the small fish to those we think might be key suspects. Everyone in this section is a possible. It will be your job to sort out who is really behind this operation and gather sufficient evidence to put him away.

"Section two: This section contains, as best we can determine, a list of all the frequent customers with pictures and bios as best we were able to develop them. We focused on the marines who frequent the bar, but added other locals when we could.

"Section three: Contains all the data on Runway 19R, the bar we purchased and where we think most of the drugs are passed. We need to confirm our suspicions. It includes the bio of the current owner, and all the documents related to the sale to you, Rick. Also, there is a listing of all the suppliers to the bar such as beer, liquor, games and food, including the names of the companies and delivery people.

"Section four: All your identification documents, cover stories for both of you, your personal bios, plus a history of each of your run-ins with the law are in this section. Rick, yours are highlighted. Examine this section carefully and let me know if either of you are uncomfortable with anything included.

"Rick, we set you up as Rick Morris, and Gary, you will operate as Gary Paul.

"Section five: The technologies we will be using, as well as weapons are included in this section. Also included are the codes we will use to communicate and the initial drop zones for information. Those drop zones will change frequently as well as the codes. You will each be given a secure military cell phone to communicate to the team. We will place surveillance cameras and sound monitors throughout the bar as soon as the current owner vacates. Gary will be responsible for keeping them operating once they are in place.

"You will find all of the technical items, weapons, your automobiles, and all your required clothing in your quarters when you arrive tonight," the colonel continued. "Rick, we have rented an apartment for you near the bar. Everything is in place in case someone checks.

"We will discuss section six tomorrow morning.

"I suggest you two spend the rest of the evening getting to know each other. The next round is on me. How well you know and develop trust in each other will be critical as the case unfolds. Your safety may depend on it!"

Looking at both of us, the colonel said, "Good evening, gentlemen, I will see you at o' six hundred tomorrow morning in my office for your next briefing." With that, Colonel Jackson proceeded to leave through the emergency exit with no alarm.

"How did he! Oh, never mind!" I said.

Gary and I spent the next two hours talking. We shared experiences, discussing the case, enjoying each other's company, while consuming several Budweisers.

I arrived at my BOQ quarters around midnight, and everything Colonel Jackson promised was there. I spent the next couple of hours rereading the briefing book and fell asleep only to awake to an early morning call from Colonel Jackson's office.

"This is your wake-up call, sir," the person on the other end of the line said, hanging up before I could respond. Colonel Jackson knew me too well!

"Sir, is there anything I can get for you?" The voice quickly brought me back to reality. "We have several snacks forward near the galley, and the coffee is hot. I believe you are my only flier awake!" The flight attendant said, standing over me. "Can't sleep on these long flights, eh?"

"Just a lot on my mind this trip. Thanks for offering, but I'm okay," I replied.

"Well, be sure and let me know if you do," she said as she slowly moved up the aisle, checking her temporary domain.

I glanced at Terri as she slept soundly in the seat next to me, cuddled up in her blanket. We made several long flights to Asia, and she has the remarkable ability to sleep soundly no matter the conditions. Me, I sleep very little! I readjusted her blanket and quickly returned my thoughts to Beaufort.

It was 0600Saturday, January 3, when I arrived in Colonel Jackson's office. Jack commented to Major Hildegard as I entered, "See, I told you he would be here on the dot! Of course, my wake-up call helped."

"Morning, Rick," Gary said, yawning because of the early hour and the same lack of sleep I experienced.

"Morning, Colonel, Gary," I replied. "Gary, you, too, could have the same wake-up service with just a little practice of being late for early morning meetings."

"Not me, I have a newborn that normally would sound the alarm. For some reason, this morning, I thought I heard him and woke up on time. Sometimes he makes sure I am awake at some of the strangest hours!" Gary replied.

The colonel set the stage. "Gentleman, Beaufort Chief of Police Phil Knollson, Special Agent Franklin Johnson of the DEA, and Major Derek Anderson, the base MP commander, will be joining us at 0700, but before they arrive, I wanted to go over a few things. First, I trust everything was in order at your quarters when you arrived there last night, and second, have you studied the briefing book?"

We both nodded in the affirmative, so the colonel continued, "I expect to begin this operation Monday morning, January 5, with you, Rick, taking over the bar. The plan is for you to meet the two employees of the bar introduce yourself and tell them the bar will be closed for a week while you do some inventory counting, maintenance repairs, and painting. During that week, we will do a full sweep of the place and install small surveillance cameras in strategic locations, as well as a voice recording system throughout the bar. We will have the capability to monitor each system from here using a secure satellite, and Gary will have the same capability at the bar using a mobile system.

"Rick, I suggest that after this meeting, you take the rest of the day to quietly visit the bar site and surrounding area. Saturday is a big day for the bar. Keep it very low level. I have arranged for you to meet with Jim Baldwin at the bar after closing tonight. This will be your one and only chance to ask him any questions you may have. He will be gone right after your meeting."

The colonel continued, "Baldwin has been under twenty-four-hour surveillance for the last two weeks to ensure he keeps his part of the bargain and to ensure we have no breakdowns in the plan. He will be sequestered in a safe house out of harm's way until this is over."

I nodded.

"Gary, best you lay low until the day you are supposed to arrive at the bar. You don't have any need for advanced knowledge since you are new in town. Best you rent a motel room in the area for the duration of this case. As indicated, you will be responsible for onsite control of all the equipment once it is in place and covering Rick's back. I have arranged a briefing for you at 1700 Sunday evening with the installation team so you can familiarize yourself with the equipment you will be using and the protocols."

Gary acknowledged the colonel as a knock on the door interrupted. "Chief Knollson and Special Agent Johnson are here to see you, sir," the days duty officer said, entering the room.

The colonel replied, "Send them in, Lieutenant, and please ask Major Anderson to join us."

Both men entered at once, as if each was trying to be the first on a crime scene! Chief Knollson fit the stereotype of a Southern police officer, white, about six feet four, a bit overweight, with sunglasses and large-brimmed trooper's hat. Special Agent Johnson was his opposite: black, small, about five feet five, slightly built, with the look of the stereotypical green-eyed shaded banker.

After introductions, we moved to the colonel's large oblong-shaped table to begin the briefing. Section six of the brief was on the table as promised, marked "Eyes Only." Major Anderson arrived just as we were sitting down and was acknowledged by the group.

Special Agent Johnson began, "Here's what we know today. We have traced the original point of distribution to New Jersey. It appears to be arriving from both Mexico and Columbia. High quality stuff that has not been cut. Our informants have told us the Giordano family is the main distributor, but we have been unable to date to gain enough evidence to charge them, but we continue to keep them under surveillance, developing information and building our case. We have tracked shipments to Charlotte through a well-known local dealer there who makes the connection to, we believe, one Dexter Cummings, a local sleazebag, that has been on our radar for some time.

"Our Mr. Cummings has been tabbed as the leader of a gang of toughs who, we believe, is also responsible for several protection shakedowns with area businesses and has a substantial small crimes record."

The chief joined in. "He hangs out at Runway 19R and appears to have been friendly with Jim Baldwin. We think both of them may have been involved with sales to marines at the base, but have not found where or who the link is to the base or if there is a link. Baldwin has consistently denied any knowledge of drugs being passed or used in his bar. He did acknowledge Cummings frequented the bar and that Cummings was not a friend. He would only acknowledge that they had a owner-customer relationship. We will interview Baldwin again, Rick, based on what you and Gary learn during your time at the bar. We believe he knows more than he is telling us."

The conversation moved to Major Andrews. "As Colonel Jackson has told you, we have developed an informant. We believe there must be one person behind pushing drugs to our marines. It doesn't make sense to have several small dealers on the base. It's logical to assume someone would organize each dealer to maintain control of distribution. We hope our informant will help lead us to the individual in charge. Corporal Liston, our informant, swears he does not know where or who is his supplier. He claims he rides his Harley on the day and time of the exchange. A cell call advises him where to make the pick-up of his next shipment, and he deposits the cash from his prior sales. He told us he has no control of the amount of cocaine he is to sell, but knows he had better sell it.

"The pick-up and drop spot is never the same, and Corporal Liston gets short notice. Up till now, we have had no time to arrange for surveillance prior to the pick-up and drop location. We have followed Corporal Liston and have investigated the drop spot, but have learned nothing of value. The drop has been made before the cell call to Corporal Liston. We wanted to mark the cash used in future drops, but were afraid we would lose the advantage of having an informant. Liston told us he had no knowledge if others are doing the same thing. He only knows his general route and routine."

Special Agent Johnson continued, "We don't want to move in on the Giordano family, nor the contact in Charlotte, until we have solid proof of the distribution chain and have arrested those involved here. Our hope would be to make arrests in each city once you and Major Hildegard have enough evidence on the leader here."

"I don't need to impress upon both of you this assignment is critical to breaking this case. We crack this case, and the rest will fall like a house of cards, but this could also be personally dangerous. We know Cummings is not above violence to protect his business," Colonel Jackson emphasized. "We will take every precaution to protect you, but keep your wits about you and don't take any unnecessary chances. This case is not worth losing two fine officers. Any questions?"

Gary and I had a few procedural questions that were answered, and we completed the meeting with handshakes and good luck wishes around the room.

The balance of my day was spent walking around Beaufort, watching who went into and out of the bar and observing activity around Runway 19R. I spent a few hours in each portion of the day and returned to the bar at 0300 early Sunday morning, my appointed time to meet with the current bar owner.

I entered the bar on time through the back door Baldwin left unlocked for me and introduced myself. We spent until 0800 hours working together, familiarizing me with the bar, its operations and everything else I could think of from my prepared notes and observations.

The bar opened each day, except Sunday, at 11:00 a.m. civilian time, so I returned to my apartment to study my notes and the briefing book. I spent Sunday afternoon watching the NFL playoffs, retired early to get a few hours of badly needed sleep.

I was standing at the back door of the bar at 10:00 a.m. on Monday, January 5, with the key in my hand, wondering if I was ready for what lie ahead.

I pushed the key into the lock and turned it. "Oh well, ready or not, here we go!"

An overpowering smell of crushed old cigarettes, stale beer, and sweat hit me squarely in the face upon opening the back door of the bar. The odor and my trepidation didn't mix well as I surveyed the main room. Runway 19R was an old bar in the middle of Shanklin Road, located near the end of Runway 19R, the main runway for the base. The original owner several owners ago had retired from the marines with his last station at MAG-31. I was told he thought the name was quite humorous and hoped it would attract marines from the base. I knew the bar from my first tour at MAG-31 and believed it succeeded because of its proximity to the base and its reputation for keeping gossip from spreading outside the bar's confines.

The bar was one big room with very old and dirty hardwood floors that were past any appearance of a shine. Years of wet mopping had made them dull. Unlike most drinking establishments, the main bar had no mirror along the back wall. A portion of the wall was fully stocked with every liquor staple, as well as a few specials that had been requested over time. Banners of platoons, squadrons, companies, and battalions of the various armed forces were draped along the wall, adding to a confusing spectrum of color, shape, and size. A very large Stars and Stripes fully lit with bright spotlights adorned the center of the wall. The flag showed wear from many years of smoke and direct light.

The bar itself was made of light oak. Unlike the floor, it shone from many years of leaning elbows, scraping of bottles, spills and gouges, and initials carved by knives of drunken patrons. The foot rails were made of brass with what were once gold-plated connectors. Both were now worn from years of use. Some of the rails actually swayed from too many years of heavy booted feet.

There were padded red leather booths along the walls that, too, were past their prime, with cracks and what appeared to be cuts in the seats and backs. Tables and chairs that occupied the center of the room were also of light oak. They followed suit with noticeable wear and tear.

The room was decorated with pictures of almost every type of military aircraft or vehicle ever used in the US military, placed there by the military customers that frequented the bar over the years. Hats, helmets, insignia, small battle flags, uniforms, even combat boots hung throughout the bar, also donated by patrons. Nothing matched, but everything looked just like it belonged.

Low-lit Tiffany lights, featuring red and yellow, hung from the ceiling above each booth and several of the tables in the center of the room, giving off small glows of light. The two pool tables at one end of the room sported the famous elongated Tiffany pool table design lights, and several small-screened televisions constantly tuned to sports were hanging on the walls.

The restrooms located behind the pool tables identified men and women as "Gents" and "Dolls" with appropriate wood-painted characters.

The large back room of the bar contained a small office and shelves for the various items of inventory, including a locked fenced section that housed the more expensive items.

The owner's office occupied one corner of the large back room. The walls inside and out were constructed of unpainted plywood, and you entered through an ill-fitting white metal door. Everything inside was definitely old government surplus. A small gray metal desk sat in the center of the room. Two metal chairs with worn black imitation leather upholstered seats faced the desk. There was one four-drawer gray filing cabinet, the same vintage as the desk and chairs, that was haphazardly stuffed with so much junk that most of the drawers hung open. The desk was a jumbled mess of paper, including old invoices, newspaper advertisements, magazines, a calculator, a telephone, and stacks unidentifiable pieces of paper. An outdated calendar hung off kilter on the wall directly behind the desk, and thumb tacks held various notes and scribbled telephone numbers. I had sorted the most important documents the night before and put them in a pile on the right hand corner of the desk. One single light bulb hung suspended from the ceiling, giving off just enough light to be passable.

Promptly at 10:30a.m., just as the previous owner said, Gladys Miller, the barmaid, and Fred Lester, the dishwasher and cook, entered the back door. Gladys, over fifty, full figured with dirty blond hair, wore a short black skirt and a tight white blouse with several buttons opened, accentuating her full breasts. I am sure the "girls," as I later found out she called them, pulled in a bevy of tips nightly.

Fred Lester looked as if he was on his last days. Extremely thin, he had wisps of what had once been black hair that hung across his forehead and sunken eyes that were set deep into their sockets. His skin was almost gray, and he was smoking an unfiltered cigarette. He was dressed in jeans and a white t-shirt. I was immediately concerned about his cooking food, but decided to let events take place prior to saying something.

"Who the hell are you, and where is Jim?" Gladys asked. She pointed her long bony finger at me.

"My name is Rick. I am Jim's cousin from Missouri, and I now own this place."

"You answered my first question, now how about the second part. What happened to Jim?"

"Jim had a little trouble with the IRS and decided it was best he quickly move on to somewhere safe. He called me and asked if wanted the place. I did, so here I am."

"Well," Gladys snorted, "just don't fuck with my customers or my tips, and we will do okay. Never really did like Jim, anyway!" She made her way behind the bar and began the task of readying it for the first customer.

Fred Lester just nodded, grunted, and headed for the kitchen, to prep for the day's menu special, not concerned one way or the other.

I stopped both of them with my planned statement on the bar closing for a week after today. I told them I would pay them for the week and fully expected to be back open by the following Wednesday. Both asked for their pay in advance. I paid them in cash at the end of the day, and they both seemed satisfied with my explanation and left the bar at the end of their work schedules after giving me telephone numbers where I could reach them to reconfirm next week's reopening.

The installation teams were to arrive at the bar at 8:00 a.m. the next morning, Tuesday, January 6, so I put three "CLOSED FOR REMODELING—REOPENING WEDNESDAY, JANUARY 14, NORMAL OPENING TIME" signs on the door and two large windows at the front of the bar. After a couple hours of looking around, I decided to call it a day and return to my apartment to study the briefing material one more time.

I returned to the bar at 6:45 a.m. Tuesday morning, January 6, to be there when the first installation team arrived at 8:00 a.m. I was surprised when the door appeared to be unlocked when I inserted the key. The door opened without me feeling I had turned the lock. A quick walk through the bar, back room, and office did not turn up anything suspicious. Everything appeared to be as I left it the afternoon before. Maybe it was just me, I thought. Maybe the door was locked prior to my inserting the key. If it was, it was probably defective. I had decided not to change locks anyway because someone with a key might be a suspect. No sense to make someone any more cautious.

The first installation team lead by Sergeant Anthony Kim arrived on schedule at 8:00 a.m. Sergeant Kim was renowned within the Marine Corps as the "Spy Wizard." His reputation was earned by his wiretapping work that lead to solving several high-profile espionage cases. I knew I was lucky to have him on my case, and I told him so when he arrived. His team was dressed as typical redneck construction workers with heavy work boots scuffed and dirt marred, multicolored flannel shirts, work jeans that displayed evidence of manual labor, and ball caps identifying their favorite chewing tobacco, sports teams, NASCAR drivers, or caps with camouflage hunting or fishing logos. Kim himself fit right in with his team with his muscular build and short crew cut.

The team's job was to execute a thorough sweep of the building, searching for possible sound bugs or cameras that could have been placed in the bar by Jim Baldwin or maybe Cummings. The team was also assigned the job of checking the structure both inside and out for other types of surveillance equipment, as well as the bar's telephones. If a device was found, it would be analyzed to see if it was active, and an attempt would be made to track back to the installer. Then the device would be cataloged and reported so we could possibly use it to filter our information back to the source that originally placed it in the bar.

The team operated at a quick but efficient pace. An hour into the sweep, Sergeant Kim asked for me to join him in the rear portion of the building next to my office. He pointed to an open trap door his team found within the secured fenced-in area. The team found it when they were moving and checking the inventory of liquor, beer, snack items, and cigarettes stored there for safety. Sergeant Kim lowered the trap door to illustrate how well it was disguised. Someone had spent a significant amount of time making sure the trap door would not be detected.

The room below was approximately ten feet by ten feet with a concrete floor. Lights were mounted in the ceiling, and stairs folded out from the base of the trap door for climbing in and out. A light switch was mounted on the left wall just at the base of the folding ladder. Looking down into the room, you could see what appeared to be small traces of white dust on the floor and walls.

"Has your team swept the room below the floor yet?" I asked.

"No, not yet. I wanted you to see it first and extend the opportunity for you to explore the room after my team completes our sweep."

"Thanks, how long do you think you will need?"

"We should be through with the room and related areas in an hour."

"Thanks, Sergeant, I will be in the main part of the bar. Let me know when your team has finished, and I can check out the room."

I contemplated calling Colonel Jackson and pushing off the next team who would be installing the video and listening devices. Finding the trap door and room plus performing a sweep of the room was an unexpected task that needed to be completed prior to the arrival of the next team; however, Sergeant Kim assured me his team would complete their task on schedule, and they did.

Sergeant Kim called for me to return to the back room and once again pointed toward the open door. "Looks like someone stored a substantial amount of cocaine down below. There are traces of it everywhere. The room appears to be only used to hold their inventory. There are no traces of anything else being stored there. We were able to collect some small samples we can use at a later date, if need be."

I climbed down the folding ladder and surveyed the room. Sergeant Kim was correct: someone had spent a substantial sum of money to protect their investment. How deeply had Jim Baldwin been involved? It would appear more than Colonel Jackson, Chief Knollson, and Special Agent Johnson knew. I made a mental note to call the colonel later that night. We certainly needed to do another interrogation of Mr. Baldwin. Obviously, he knew more than he told us prior to his departure.

Sergeant Kim was waiting for me when I reached the trap door opening on my climb up. "Whoever is involved must have been here recently, because we also found traces of cocaine leading to and around the back door."

"When I arrived here this morning, the door pushed open as though it wasn't locked. And I was sure I locked it last night. I thought it was probably a defective lock. Looks like someone removed the cocaine last night right under my nose. Any traces of cocaine outside the back door?" I asked Sergeant Kim.

"Yes, there are also a few scuffed-up footprints and some tire tread marks. Not deep enough tread marks to identify the tire."

"So, when they realized Baldwin was gone, someone was obviously extremely concerned about the cocaine being found, and something happening and someone decided to move his or her inventory immediately. Any idea how much cocaine was stored down there? Any fingerprints anywhere in the area?"

"No, no prints that are complete or clear enough to help make an identification, and it is hard to determine just how much cocaine was stored. You can bet it was no little stash! They were pros and left the site pretty clean."

On Thursday, January 8, Sergeant Kim declared the operation a success and told his team to wrap and prepare to leave. The next team would arrive on Friday morning.

Sergeant Pete Ench arrived with his electronics team at 10:00 a.m. on Friday, January 9, right on schedule. Their panel trucks were labeled "Charleston Electric & Maintenance." We had selected that name rather than a name from the Beaufort area to make tracing the trucks difficult. If someone did call the number on the side of the trucks, a very firm but pleasant voice would inform the caller a call back would be necessary as all the trucks were on job sites, or to leave a call back number and someone would return the call as soon as possible. The sergeant's team was tasked to install all the video and sound equipment in the locations we decided were the most critical for gathering information within and outside the immediate surroundings of the bar.

The installation went smoothly, and the team completed their assignment ahead of schedule, just prior to 4:00 p.m. on Saturday, January 10. The painting crew was not due until Monday, January 12, so I locked up, went back to my apartment to change clothes, and headed for a dinner meeting with Major Hildegard in Savannah. We had planned to meet at Bistro Savannah for dinner to discuss his training with the electronics team at an offsite secure location and what I had learned working with the surveillance teams over the past four days.

It was a pleasant but cool January evening as I made my way over the Savannah River bridge and into the city toward Bistro Savannah, located at 309 Congress Street, five blocks off the Savannah River in the historic district section just off city market.

Savannah was very different from Charleston, its neighbor to the north. Charleston was located on a large bay that leads to the Atlantic Ocean. The American Civil War began on April 12, 1861, when Brigadier General Beauregard of the Confederate Army who fired on Union-held Fort Sumter, located at the entrance of Charleston Harbor. Months before, cadets from the Citadel had fired on a federal ship entering Charleston Harbor, causing minimal damage but escalating the hostilities. The Union laid siege to Charleston for one year early in the war, using captured Fort Sumter to shell the city. The Union cannons fired long-range fifty-pound balls across the bay to the grand houses along the battery, almost destroying them and the rest of the city with deadly fires that ravaged day and night during the shelling. The blockade of Charleston by the Union Army prevented supplies from reaching the citizens within the city. The main source of food for Charleston residents came from blockade runners and from local gardens producing required staples.

Savannah, known as America's first planned city, on the other hand, was located inland on the Savannah River and had also played a deciding role in the war. General Sherman had captured the city on December 22, 1864, as part of his march to the sea. The original layout of the city had been designed by its founder, General James Oglethorpe, around 1733, with squares featuring monuments to prominent city figures and gardens featuring plants and flowers sporting bright colors during each of the growing seasons of the southern region of the United States.

I made sure I was not followed as I made my way south.

Bistro Savannah was one of Terri's favorite restaurants outside of Charleston. I entered the large windowed antique front door of the restaurant and spotted Gary sitting on the left side at one of the tables for two along the exposed gray brick wall of the former cotton warehouse. The high ceilings and pine floors gave the restaurant a unique feeling.

Gary was waiting for me when I arrived. Two Budweisers were sitting on the table in anticipation.

"Good timing," I said. "How long you been here?"

"Not long. Just long enough to peruse the menu and drink my first Bud."

"Any trouble finding the place?" I asked.

"No, thank God for GPS!"

"You are right. What did we ever do without them?"

Gary laughed. "Continue to get lost and drive our wives crazy by not asking for directions."

"You are so right. How did your day go?"

"The training was outstanding," Gary said. "I will have the capability to monitor real time both the video and audio. The technical team has developed a small monitoring unit that can be placed under the bar. It looks like a case of liquor bottles."

"Pretty cool James Bond stuff."

"How did it go with the installation?" Gary asked.

"The installation went as scheduled; however, the team discovered a trap door and a ten-by-ten room below in the back area of the bar across from the office. It has been the storage location for either Baldwin's or Cummings's cocaine inventory. You know, I searched that bar thoroughly and didn't find the trap door, so it was well disguised. The room was empty when the team found it. Someone, maybe Cummings, beat us to the punch and removed all the cocaine, probably the night before the team arrived. The team found small traces of cocaine in the room, along a path to the outside door and on the ground outside the door. There were tire marks where they apparently loaded the cocaine, but not enough to get a reliable sample."

Gary reviewed the menu. He asked, "I assume you have been here before?"

"Oh yeah, this is Terri's favorite restaurant outside of Charleston."

The waiter waited patiently while I continued, "We had our first weekend date in Savannah staying at the Gastonia Inn. The innkeeper

34

recommended Bistro Savannah, and we loved it. If it's not dinner in Charleston, it's here!"

Gary nodded to the waiter, indicating we were ready to place our wine selection and our entrée orders. I decided on my favorite, the seared jumbo scallops over fresh crab succotash. Why try something new when you have a proven favorite?

Gary ordered Terri's favorite, upon my recommendation, after confessing he was a spicy food fan. The Thai-spiced mussels with lemongrass, coconut, and red curry sauce was a signature dish of the restaurant and a ten on the spicy meter.

The Mondovi 1998 select chardonnay that arrived at the table was sampled and deemed extremely drinkable. Dinner was up to Bistro Savanna's quality standards. Gary enjoyed his mussels and agreed the ten on the spicy meter was correct!

We continued our discussions on the plan ahead and left the restaurant about 10:00 p.m. I headed back to my apartment near the bar. Gary followed to ensure no one was trailing us.

The paint team arrived on schedule Monday, January 12. The painters were marines stationed at Marine Corps Air Station Cherry Point in North Carolina and were flown to protect the secrecy of the operation. The marines doing the painting enjoyed the break from routine duty and completed a respectable job early Monday evening. A painting professional would have approved of the workman-like job of covering all the marks, nicks, and gouges that occurred during the other steps of the installation.

I spent Tuesday, January 13, taking inventory, cleaning the bar, and working up a new food menu. I had to do everything possible to convince everyone I was truly the cousin and the new owner. Selling myself and Gary was instrumental to the operation.

On Wednesday morning, January 14, I was standing at the front door at 10:58, ready to unlock the door and welcome the first customer. Gladys and Lester were busy with the last-minute routine associated with the daily opening of the bar. This was it! We had done all the right things in preparation, and I felt confident but nervous. This was my first attempt at solving a case while working undercover. My hands were shaking so hard I dropped the keys to the front door and looked back to make sure no one noticed.

Our first customer was standing outside waiting and, upon entering and proceeding to his favorite bar stool, asked, "Where's Jim?" Minus the comment about the IRS, I repeated my Gladys story. Over the next three days, I wished I had a tape recorder to play the same story over and over. Finally, most of our regulars were satisfied and quit giving me those looks saved for a petty thief or mooching relatives.

Gladys appreciated the tips I pushed her way, and Lester said little but did do a reasonable job of cooking and cleaning dishes.

A week went by with nothing out of the ordinary happening, and my first informational contact with the team was uneventful. Gary Hildegard arrived on schedule the following Wednesday, January 21, looking for work as planned. I hired him as the bartender and introduced him to Gladys and Fred. Gladys gave Gary the same opening speech she did to me, and Fred just nodded, extending his hand.

A very large man with an entourage entered the bar at approximately 9:30 p.m. on Thursday, January 22, the second week of my being in place. The room was full of off-duty marines and locals playing pool and watching the Thursday night ESPN college basketball game between the South Carolina Gamecocks and Alabama's Crimson Tide. The Gamecocks were losing badly, and my Clemson brain was happy, but only to myself. The crowd was definitely full of Gamecock supporters.

The big guy located me talking to a customer, approached, and motioned for me to follow him to a booth in the rear of the bar. I noticed, as we were walking to the back, two of his entourage not so gently removed three people sitting in that particular booth.

We sat down facing each other, and Cummings motioned for his guys to get lost.

"I'm Dexter Cummings; my friends call me 'Dog.' Only time will tell if you will be a friend. I understand you own this place now."

"I do. What can I do for you, Cummings?"

"That's what I like: a man who gets right to it," he said. His voice carried a laugh. "I have been watching you for the past two weeks, trying to figure out your angle. You see, your cousin and I had an arrangement, and just like that, he disappears. No thank you, no goodbye, and not a word about his cousin from Missouri taking over the bar. Don't you think that's strange? Just up and leaving like that!"

"Hey, look, whatever you had going with my cousin is your problem and has nothing to do with me! Jim's troubles are his troubles. He gave me an opportunity and I took it. No questions asked, no answers given. If you have a problem with Jim, I suggest you find him and settle it."

"Oh, I will if I need to, but how do I know you are really Jim's cousin from Missouri? How do I know you're not a cop or a fed sent here to screw with me? I thought I could trust Jim; now, I am not so sure. Out of the blue, you arrive, so why should I now trust you!"

"I don't give a shit if you trust me or not! I bought this bar, and I intend to run it any way I want," I said as I slid out of the booth. "Our meeting is over, Cummings, and next time, don't mess with the paying customers!" I said. I walked back toward the bar.

"Don't be so hasty, "Cummings said. He pointed to one of his boys. "Harry, please escort my friend back here so we can continue our talk."

Harry made a move to grab my arm, but I hit him squarely in the gut before he could get a grip, spun him around, and wrenched his arm behind his back, hopefully breaking it!

Harry screamed, fell to the floor clinching his arm in agony. "Now, asshole, take this loser, the rest of your hangers-on, and get the hell out of my bar!"

"As I said, don't be so hasty. Please, sit down, and let's finish our conversation. I can assure you, I know who you are, and I can find Jim if I need to, but he is not my problem right now. If you will listen, I believe we can continue the arrangement as before, and I can make you a rich man!"

"I have heard statements like that before from people tougher than you, and if you did check me out, as you say, you know I'm not rich. And by the way, let's get one thing straight: this is my bar now. I bought it fair and square, and I will run it as I choose. Don't try to muscle me or tell me what to do, understand?"

"Oh, I checked you out, as well as your new bartender over there, and if I had any questions, you and this bar would be history! I have too much to lose to take any chances. So, are you interested in making some money?"

Cummings's comments meant our cover story was effective and in place for both Gary and me. I silently thanked Colonel Jackson and the team. I motioned to Gary. "Gary, please bring my friend and me a bourbon."

Gary was walking back to his position when the door to the bar opened and Chief of Police Knollson entered. The place fell silent as some customers just stared; others tried to become very small, praying the chief was not looking for them. A couple of customers quickly departed the bar, brushing past the chief on their way out.

Chief Knollson spoke loudly said, "I understand this bar is now owned by Rick Morris. Where is he?"

Gladys pointed at me. "He's over there, Chief. The booth at the back."

I stood up from my seat in the booth, walked directly toward Chief Knollson, extending my hand. "I'm Rick Morris, and I'm the new owner."

The chief refused my outstretched hand, did not introduce himself, and with an official police voice said, "Look, Morris, I got word about you, and I don't like it! Your track record is not acceptable in this town. We like our people to enjoy a beer now and then, but we don't need any extracurricular activity that is not within the law. Understand?"

I dropped my hand and surveyed the room before responding. I wanted to make sure Cummings was listening.

"Chief, this bar is a new start for me, and I don't plan on anything but serving a few drinks and some food while trying to make a living. There is no need for a lecture!"

"Don't get smart, Morris. Just keep your nose clean so I don't have a need to return here for anything but a beer and to sample some of your food. Okay?"

"No problem, Chief. You are welcome here any time, on the house!"

Chief Knollson spun on his heels and left the bar. My customers sighed with relief and returned to their drinks and food, reacting as though what just happened was not unusual.

I returned to the booth where Cummings was fuming, "Who does that asshole think he is? What makes him think he can walk in here like a high and mighty person and talk to you like that—what a piece of shit!"

"Aw, not a big deal. He's just doing his job. I would sure like to know who ratted me out, though."

Cummings scoffed, "He'll get his one of these days! Now, where were we?"

"Oh yeah, I was about to listen," I said. I slid the shot of bourbon Gary had delivered across the table.

Not ready to let it drop, Cummings said, "I still say he's an asshole!"

He began, "You see, I run a small distribution business that supplies local customers and marines on the base. Your bar is my office and inventory location.

"All I need you to do is keep my inventory safe and keep the cops off my back. For that, I will give you a very reasonable cut."

"Yeah, a small distribution business, my ass! The only thing people distribute from a bar is drugs or bootlegged liquor, and this is certainly not a dry county." Cummings did not respond to the opening I gave him, just stared at me. I needed him to admit he was dealing drugs, but would have to press that issue in further discussions. "Look, this bar is not a bad business. With the proper care, it can make me a living. You heard the chief. I don't need to give him any reason to slam me in jail. I'm not going there for you or anyone else. Also, I don't do drugs, but to each his own. It's not my job to act as society's watch dog!

"How do you expect me to keep the cops out?" I asked.

"Don't worry about the locals. I have them covered," Cummings said. "A frequent little token of my appreciation is enough to keep them in line, but the feds are another story."

"How will I know which locals, and how much do you pay them? There are a bunch of cops in Beaufort. What about the base MPs or the state boys?"

"You just let me worry about who and what it costs me. It's not coming out of your cut. My guys keep the rest of the local cops out of here unless they are with them. The state troopers couldn't care less about what happens in this dump. The MPs only come when someone calls them, and nobody does.

"I haven't had fed problems as of now and don't want any," Cummings said. "You know you can smell a fed a mile away, so when you smell one, make sure to turn his attention somewhere else!"

"You mentioned supplying the base. How does that work? You must have contacts there."

"You ask too many fucking questions. Just do what I asked; keep your damn mouth shut and your eyes open. You do that, and we will get along just fine, and you'll make some very easy money!"

"You said your inventory is kept here in the bar. Why here in the bar? I'm sure you have much safer places where you can store your inventory. This bar really does not seem to be a safe place to me."

"Would you look in this bar if you were looking for drugs? I never allow any drugs to be sold here so there is no interest. No rumors or meddling. This is the last place the cops would look. It's totally safe, and I have you to watch it."

"Okay, so where's the inventory kept?"

"There is a trap door just inside the backdoor on the right side," Cummings said. He pointed to the back section of the bar. "We move it after hours for security. All you have to do is be here, do an inventory verification, and make sure there are no interruptions or nosey little faces around!"

I mentally thanked the team for leaving the trap door and storage room just as they found it and said to Cummings, "What trap door in the back room? I have checked this place out more than once and didn't find any trap door."

"The door was built to be undetectable. You will see it in due time."

"So, what do I get for taking these kinds of chances?" I asked.

"Your cousin's cut was a generous 5 percent."

"You're joking, right! You must think you're dealing with an idiot! For 5 percent, my answer is fuck you. Now get your shit out of my building before I do something you won't like!"

Cummings sat there and just looked at me for several long seconds. I was not sure if I had pushed too far, but it was critical for Cummings to respect me.

Before he spoke I said, "I want 50 percent."

Cummings glared at me, "Twenty percent, you greedy bastard, and that's it. Don't push me, or I might do something you won't like!"

"Deal, but no individual selling in the bar, and I want the names of everyone involved in the distribution. Where's the crap coming from and how do I know I can trust your source?"

Cummings replied, "Where and who is not your problem. You just keep up your end of the deal, and we will be okay."

"What about Gladys and Jeff? Are they involved?"

"Fuck no, and you keep it that way, understand!"

I nodded.

"When can I get my list?"

"When you need it, but for now, keep my inventory safe and your eyes open."

I felt I had pushed enough during the conversation with Cummings and hoped the cameras and video were working. If so, we had the beginning of a case against him. We had enough to charge him with bribing a police officer, but that was not the issue.

I excused myself to return to the bar and the growing number of afternoon customers. It was time to arrange a meeting with my contact and report today's happenings. Cummings left the bar, promising to return later that night.

I waited an hour before making a call on my secure military cell phone to my contact, Major Derek Anderson, establishing a meeting with the team early Friday morning, January 23, at 3:00 a.m. at the pre-identified location.

The 3:00 a.m. meeting took place on schedule. I met Major Anderson, Chief Knollson, and Special Agent Johnson at the appointed location after driving a circuitous route, making sure I was not followed. Gary followed discretely behind me as a double check.

Both Chief Knollson and Agent Johnson had received the briefing from the installation teams and agreed everything was in place to record events surrounding Cummings and the distribution of drugs.

I relayed the story of the trap door and storage room conversation with Cummings and my suspicion that he had emptied the room of the evidence the night before the teams arrived and my incident with the backdoor key.

All three listened intently as I relayed my conversation with Dexter Cummings. Johnson was the first to speak after I finished my report. "We have a tail on Cummings, and he was in Charlotte the past two days, meeting with his distribution contact. Obviously, his checking and your story rang true to him, or he would not have made the trip to Charlotte and placed an order. Unfortunately, there was a glitch in the software, and your conversation today is not on video or sound. The glitch has definitely been repaired. Sorry!"

"How in the hell did that happen? I thought you had arranged for the world's best teams to do the installation!" I said. I meant my tone to be sarcastic.

Johnson replied somewhat defensively, "They are the best! This time, I have personally checked everything myself and can assure you everything is working! Now, we need you to get him to say the same things now that we have everything under control."

"It had better be," I said. "I can't do this too many times before Cummings gets suspicious and the whole thing is blown, maybe including me!"

Chief Knollson followed, "I also need the names of the local police officers involved so we can keep an eye on them. We need to find out how Cummings gets the dope onto the base. He must have direct contact with someone as a coordinator. Did he mention anything about how he does the distribution?"

"No," I said, "and I don't want to push too hard and scare him off. I'll work on that next week."

"Corporal Liston has agreed to work with us," Major Anderson continued. "As of now, he knows nothing about you or Gary. He will be released from the stockade tomorrow and will find out about the bar ownership change on his next visit there. We need to verify he is telling us the truth about his part in the distribution. Thanks for your visit today, Chief. You were very convincing and Cummings bought it. He sure hates your ass! Gary was on top of it and called me as soon as that prick got to the bar. Cummings should hate me. He knows I will put his ass away the first chance I get. This is not my first rodeo with him!"

Nothing out of the ordinary happened the rest of Friday and Saturday, day or night, and Cummings was nowhere to be seen. I felt certain he was testing me to see what I would do after our discussion. I also felt comfortable the proper tone had been set with my approach to him and the visit from Chief Knollson. There was no doubt in my or Gary's minds that Cummings was a dangerous character and that we needed to be extremely cautious in our dealings with him.

The weather had changed in Beaufort from a very late Indian summer to winter almost overnight. It was Carolina cold when I opened the bar the following Monday, January 26. I looked for Cummings to arrive early that day, take up residence in his booth of operation, and begin making deals for the inventory. Cummings would need to replace what he had removed and add inventory for the expected incoming orders.

Cummings arrived on schedule, and activity heightened at his booth office. Several individuals met with him during the day, and it became obvious Cummings was assuring each potential customer that he was back in business. I did not recognize any of the individuals as frequent customers. I guessed they only came to the bar to do business, not drink.

About 4:00 p.m. that day, Monday, Cummings approached me at the bar and asked to meet in my office. "We will be bringing in more inventory tonight at three a.m. I need you to be there for the counts."

I agreed and asked Cummings, "Where is the inventory coming from?"

"From Santa Claus, asshole. Just be there and keep your mouth shut like I said."

Two local police detectives wandered in around 4:30 p.m., ate some of my food on the house. They spent a few minutes with Cummings. He passed an envelope to each detective. They quietly thanked him and left the bar.

I asked Gladys who they were, and she identified them as Beaufort police detectives, William Miller and Buck Thomas. As she put it, "Those two are just scum! I wouldn't even trust my dog with them. Those motherfuckers would slit their grandmothers' throats for a buck!"

I made a mental note of the names and descriptions to send along to Chief Knollson and Special Agent Johnson.

Corporal Liston arrived about3:00 p.m. on Monday, January 26. I recognized him by his picture. He had a few beers with the person who came with him, who was not on our customer list. Liston didn't appear to know Cummings and left the bar with his friend about 5:00 p.m.

Gary and I met in his motel room later that evening to review his pictures and the video and sound recordings gathered that day from the miniature surveillance cameras. This data would be used as evidence when the case was complete and went to trial. This time, all the equipment worked as planned. The software glitch appeared to be a thing of the past.

I arranged to drop off the latest information to the team at the pre-arranged drop point.

Gary and I discussed how to get more information from Cummings on his local distribution contacts and his supplier. It was critical to develop this information without arousing any suspicion.

We also discussed potential options for getting information about the identity of the possible leader distributing drugs on the base. We had no clues at this time about his identity but hoped between Cummings and Corporal Liston there would be a breakthrough.

It was barely raining when I arrived at the back of the bar at 2:45 a.m. early Tuesday morning, January 27. It had been raining hard all day, but had slacked just at the right time. Gary and I were still uncomfortable with the surveillance equipment, so Gary double checked everything to assure it was functioning correctly.

Cummings arrived with three of his guys at 3:00 a.m. They were in a rented white Ford delivery panel truck.

"Door unlocked?" Cummings said. He pointed to the back door.

"No, I just got here." I unlocked the door and turned on the lights in the back room area. The lights gave off an eerie glow in the falling rain turned to mist.

Cummings directed one of his guys to watch the back parking lot and for the other two to begin unloading the bags of cocaine. He entered the back room, opened the door to the fenced area. I helped him move the boxes of bar supplies that were covering the trap door.

I pretended not to see the outline of the door as Cummings commented, "See, asshole, there is a door just like I said. Made it special to keep my stuff safe."

It took the four of us about a half hour to unload and stack the bags in the room below the trap door. I counted each bag, and Cummings and I agreed with the count. "There had better be the same count when I make my first withdrawal, or you're fucked, understand?"

"Don't worry; like I said, I don't want anything to do with how you handle this garbage. You make sure you include me when you make each withdrawal, and we will get along just fine."

By 4:00 a.m. Tuesday morning, we completed the transfer, and I headed to my apartment to get a few hours of sleep prior to opening the bar.

Gary and I had our first opportunity to review the video and audio later Tuesday evening, drinking a beer right after closing time. This time, everything worked as designed.

"Well, we have our first piece of evidence," Gary said. He shut the equipment down.

"Yes, now we need verify where the dope is coming from and who is involved at the base."

"And do it before Cummings figures out what we are up to!"

"Let's get out of here. Tomorrow could be a very interesting day."

Cummings arrived earlier than usual on Wednesday morning, January 28, and business began almost before he was comfortably settled in his booth office. I did not recognize any of his customers, and true to his word, no cash or drugs were passed. Corporal Liston did not make an appearance, so I began to believe his story about the transfer of cash and drugs. Obviously, word was out that Cummings once again had inventory, and he was open for business.

About 4:00 p.m., Cummings got my attention and motioned me to his booth. "We need to make an inventory withdrawal tonight. Be back here at three a.m."

I agreed and went back to my customers. Gary and I discussed the day's happenings after I closed the bar, and he made a check of the equipment in the back room prior to our leaving.

Once again, I arrived about 2:45 a.m. on Thursday, January 29, and waited for Cummings and his guys. They arrived on schedule at 3:00 a.m. driving another rented delivery van. This time, they sorted specific quantities of cocaine from the inventory, made sure I agreed with the counts, and began to leave. I said to Cummings, "Where are you taking this stuff? It's obvious the bags are for specific customers. How do you deliver the goods and get your money?"

"Look, shithead, like I told you, your only problem is to keep my inventory safe and correct. You don't need to know anything more!"

"How do I know how much my cut is if I don't know how much this stuff will bring when you deliver it and the details of each deal?"

I continued, "When are you going to show me the books and give me my cut? I don't know how you did it with my cousin, but you're not going to fuck me like you probably did him. You're going to show me, or the deal is off!"

"Listen, fuck face, you will get your cut when the deals are done and I have my cash, and as far as your cousin and my fucking him, well, let's just say where he is, he won't be worried about any of this anymore.

"Now lock this place up and get the hell out of here."

I was immediately concerned about the comment about Jim Baldwin but felt I had pushed as hard as I dared. I watched as Cummings and his boys left the parking lot and headed east on Shanklin Street.

Special Agent Johnson and his team were waiting down the block as we had discussed, and they began a tail as the van passed their location. An elaborate surveillance plan was in motion to follow the van to its final destination. We assumed Cummings would meet his customers somewhere and distribute the cocaine. The team would act if the leader of the drug ring on the base was clearly identified. If not, the team would split up and follow individual customers but make no arrests. Each team would video the person they were following for later identification and data for any future trial.

I left the scene and returned to my apartment. I slept very little that night, worrying about the surveillance operation and how well the equipment had functioned.

I received a secure cell phone call Friday, late in the morning of January 30, from Major Derek Anderson, my contact, requesting a meeting with Gary and me at 2:00 a.m. on February 1 at the next appointed location. I agreed and notified Major Hildegard. We were both anxious and apprehensive about the events that occurred once Cummings left the bar early Thursday morning.

Gary and I arrived at the meeting location on time to meet with the team. The news regarding the day's events was mixed. The surveillance team followed Cummings and his van to an old, small, dilapidated warehouse downtown where the cocaine was divided between Cummings and each of his gang, with Cummings loading the largest portion into his gold Cadillac Escalade. There were too many carriers for the team to follow all nine, so one team followed Cummings and the rest of the team followed five of the gang.

"We did obtain photographs, video, and some audio on the individuals we were able to follow as they made drops and collections, and also we were able to hold a secure position at the drop point where we photographed, videoed, and recorded audio of each pick-up. Each individual picking up was different than the individual making the cash drop. We are not sure at this time if both individuals were acting as a team or if the quantities were equal for each drop. Unfortunately, all the individuals we saw were civilians. We did not identify any marines or anyone associated with the base at any of the drop sites. Corporal Liston was not involved. That leads us to believe Liston gets his directions from whoever is the leader on the base.

"Unfortunately, Cummings gave us the slip, so we did not see where or whom he made his drop to. We had our best team on Cummings, but unfortunately, we believe he made us. We broke off at the first indication he was aware he was being followed. If so, you guys watch your backs. He may become suspicious of you, seeing as you are the only new players in this thing. Cummings is street smart enough to know the tail had to start at the bar. I also believe he knows it was not the local authorities."

"There must be something else going on. I pushed him on how my cut was handled and how did I know I was getting my fair share. Then I told him he was not going to fuck me like I was sure he did my cousin. Cummings's reply was my cousin would not be worried about any of this anymore. Not sure what he meant, but it can't be good."

Special Agent Johnson replied, "We had planned to re-interview Mr. Baldwin this week anyway. Looks like we need to make that a priority! We have him well hidden until this is over. I don't believe Cummings can locate him."

Gary and I left the meeting concerned we were no further ahead on identifying the base contact, but feeling good we were developing a solid case on the rest of the gang. All we had to do now was find a way to identify the base ringleader and get the goods on him. That was not going to be easy if Cummings did make the surveillance team and became suspicious of us. Gary and I would need to double our efforts to protect ourselves and keep Cummings convinced we were legit.

Cummings spent very little time at the bar the next day, but acted normal. I asked him when I could expect my cut from last night's operation and was told it would be soon.

The rest of Saturday was normal for the bar. Customers arriving after work were ready for a weekend of college basketball and Sunday's NFL Super Bowl. I had planned to be open on Sunday, February 2, and arranged for a Super Bowl party at the bar with special drinks and food, with prizes for the best team costume. If this was truly a money-making venture, Super Bowl Sunday was the bar's largest cash intake of the year.

Gary received a secure call from Major Anderson around 5:00 p.m. with very disturbing news. Jim Baldwin was found dead in the safe house located in upstate Maryland that afternoon along with four DEA agents.

"My God, Gary, Cummings told me what he was going to do," I said, reacting to the tragic event. "I heard but didn't listen. I could have warned the team. If I had said something, Baldwin and four good men would still be here today!"

"Come on, Rick; don't beat yourself up on this one. How would you have known Cummings meant he was going to have Baldwin killed?"

"I just thought he meant Baldwin was gone, not a part of the deal anymore; therefore, he would not care what was happening."

The four members of the security team were dead from gunshot wounds. The fifth team member was left for dead but had survived. He was in critical condition and unable to communicate.

Gary said the attackers had disabled the security devices around the property and had arrived in a delivery van posing as a person delivering food that had been expected that Sunday afternoon. The agent on station at the gated entrance to the house had been killed instantly and the gate opened from inside the guard position.

It was believed there were five attackers, all carrying AK-47 assault rifles, Micro UZIs, and 9mm Berettas. A concussion grenade had been thrown into the lower level of the house, and the front door was blown at the same time. A second agent was shot down as he attempted to return fire from the kitchen area. He survived by crawling inside a kitchen cabinet.

The remaining three agents attempted to remove Baldwin down a back staircase, but were met with gunfire and retreated to an adjacent bedroom. One agent did manage to use his cell to call for backup, but everything was over when backup agents arrived. The remaining agents and Baldwin never left the bedroom and were found in various defensive positions in the floor and on the bed.

The attacker carrying the UZI must have fired aimlessly because there were bullet holes in every room of the house, especially the bedroom where the three agents and Baldwin were found.

It was not clear who was responsible for the attack and killings, but I believed Cummings was the key suspect. Obviously, he did not lead the attackers but was the only person that had anything to gain from Baldwin's death.

Major Anderson requested Gary and I meet the team at the next identified location at 3:00 a.m. Monday morning and to keep clear of Cummings outside the bar. It was imperative a solid plan was in place to not only protect Gary and me, but to secure the operation. Today's events made catching Cummings and the rest of the chain critical. The case was now more than just a drug bust.

"Now, I really want this bastard," I said. "He must be stopped and put away for good before anything else horrible can happen!"

We went through the usual bar-closing procedure, and the bar closed on schedule at 1:00 a.m., Monday, February 3, after the crowd dispersed after the Super Bowl game.

I retired to my office to finish paying some invoices and kill time until our arranged meeting time with the team.

I was playing the day's events back in my head, lamenting the senseless killing of four good men just doing their duty and wishing I had been more alert. I was also developing a mental plan for dealing with Cummings. The case had reached its most dangerous point. Obviously, Cummings knew he had been followed and for some reason felt he had to eliminate Baldwin. He had as much as told me what was going to happen, and I should have been more cautious. I had to convince him that Gary and I had nothing to do with the events of the past two days. It was also imperative Cummings not know I knew anything about Jim Baldwin.

I had just looked at my watch. It was 2:00 a.m. I should be leaving soon. The rendezvous point was half an hour away, but I needed extra time to assure I was not followed. I thought I heard the door back door opening and cautiously walked from behind my desk to check the noise. Cummings appeared just as I reached the door of my office. I knew immediately this was not a social visit and mentally scanned the office for any type weapon. Cummings was on me so fast I had no chance to reach for one anyway!

He grabbed me, spun me around, and placed his forearm on my throat. "I told you I would tell you when you were my friend; well, asshole, it looks as though you are not going to make it!"

He threw me across the desk, and I hit the office chair so violently I cracked my ribs. Blood was coming from where my mouth hit the corner of the desk and a few teeth would need repair if I made it out of this one. Cummings threw one of the surveillance cameras at me. It hit me on the top of the head, causing a two-inch gash.

"Did you think I wouldn't find that shit!" He pointed to the camera. "How stupid do you think I am?"

I made a feeble attempt to stand while attempting to formulate a smartass answer!

Cummings shoved me again, and I stumbled back into the desk chair. "So, who are you, a fed sent here to get the goods on me and my operation?

"You fucker, you're sure as hell not Baldwin's cousin, "Comings continued before I could answer. "He was kind enough to tell my boys that right before they put a bullet in his head. First he swore he did not know who you were, but changed his tune after a not so gentle persuasion. Stupid bastard really thought we would let him live if he told us the truth. Now it's your turn, motherfucker. Just who are you?"

Once again, I staggered up from the desk chair, quickly surveying any possible escape route. Even if I made it past Cummings, his guys were just beyond the door!

I just looked at Cummings and refused to respond.

Cummings grabbed me and dragged me across the floor toward the front door, kicking me in my side along the way. The pain was excruciating! I felt myself blacking out but fought to keep awake. I had to do whatever it took to stall for time.

I knew the team would quickly realize something was amiss when I didn't show and begin a search for me.

"Don't you go away on me, you son of a bitch!" Cummings shouted. He noticed my state of consciousness. "You and I are going on a little trip. I have someone who wants to meet you and discuss the situation you're in! It's not pretty, is it? You dumb fuck, you're as stupid as your so-called cousin. My friend and I will teach you or anyone else who takes your place not to fuck up our operation!"

We reached the door and Cummings hissed, "Now, stand up and get in the back of that van." I recognized the white van Cummings used to deliver the cocaine earlier in the week. A second van parked next to the white van.

"Jerry and Randy, you two come with me. Mickey, you take the rest of the guys and load the coke into the van. After you are done, take it to the warehouse and wait for me. Your asses had better be there with my coke when I get back, or I will find each of you and cut your balls off." Cummings pushed me in the back so hard I fell, nearly hitting the bumper of the van. I managed to brace myself with my right hand and struggled to regain my balance.

"Get in the back, asshole," Cummings said as he threw me into the back of the van. I landed hard on my cracked ribs and moaned. "What's the matter, Morris, or whatever your name is? I thought all you fed pricks were tough guys trained for just this shit."

"I'm not a fed and have no idea what you are talking about, "I said. I tried to get my bearings and decipher my best opportunity for escape, realizing now was not the optimum time. Even if I could overpower Cummings, someone would shoot me before I could get control of his gun and return fire.

No, I would just have to wait and see where we were going. There had to be an opportunity!

Cummings crawled into the van and stuck his 9mm against my head. "I would put a bullet in your head right now if I could. Now, where is your partner, the bartender?"

"I don't have a partner. Gary was just a guy needing a job, and I needed a bartender. Leave him out of this! He knows nothing."

Cummings smashed his revolver across the top of my nose, breaking it, making me nauseous and dizzy. Shit, that hurt! Bastard! Blood squirted down the front of my shirt. I used my sleeve to wipe away as much of the blood as I could and waited for the bleeding to stop.

"Don't lie to me, you son of a bitch. It's too late for that."

I started to respond, but Cummings cut me off with a slap across my face!

"It makes no difference anyway, asshole. My guys will find him, and believe me, they won't treat him as nicely as I have treated you, you son of a bitch!"

"I told you, Gary had nothing to do with this. Leave him alone! You do anything to him, and I will make sure you die a slow, fucking torturous death."

"You talk big for someone with a gun to his head and no way to stop me! In fact, I wish you would try to jump me—then I could shoot your sorry ass!" Cummings said. He giving me another love tap on the head, once again almost sending me over the edge to blackout city!

"Jerry, let's get out of here. You know where we are going. Keep it nice and slow; we don't need any trouble from a nosy cop!"

The van lunged forward out of the parking lot. I couldn't see where we were headed, but by the movement of the van and my estimate of time, I knew we had left the city streets about thirty minutes ago. The van made a sharp right, and the smell told me we were somewhere around saltwater.

"You don't really think you will get away with killing me, do you? Your operation is finished, and there is enough evidence to put you away for a long time. My guys will find you and hang your ass!"

"I am going to enjoy seeing you die, you bastard! It took me years to get this business running like clockwork, and then you come along and fuck it up. Once you are out of the way, my friend and I will just move on to another location in Beaufort. Neither you nor the feds are going to stop us. You see, my friend will make sure all that evidence, as you call it, will just disappear, and so will you!

"Even if your bartender friend is not part of this, as you say, nobody will miss a drifter like him anyway."

I was beginning to regain awareness of things around me. My ribs were killing me and breathing was difficult, but I knew I had to prepare myself to take advantage of any opportunity to escape. There had to be an opportunity! I had a feeling wherever we were going might be my last chance to survive.

The van rolled to a stop, and we waited until Jerry opened the back door of the van. Cummings pulled me out the back door. I fell on the ground, landing face first. My nose began to bleed again.

51

"Now look what you've done, asshole. You are not going to bleed all over my boat!" Cummings bent down, put his 9mm to my head for the second time, and said, laughing wildly, "Maybe I will just kill your raggedy ass right now and save all this! Bang, bang, asshole! Now, get up and clean up your nose."

I recognized generally where we were when we arrived at the top of a very long landscaped sand and seashell driveway that lead to a southern-type mansion situated not far from the entrance to Port Royal Sound. A pine forest was to my left as I stood up. It would be a short boat ride from the house on the inlet where the dock was positioned to Port Royal Sound and the ocean. I assumed this was Cummings's home. If so, drug smuggling was a very profitable business! The four of us walked slowly toward a dock behind the house. At the end of the dock sat my boat! Well, not really my boat, but my dream boat. Athirty-six-foot1981 Grand Banks Classic sporting a Lehman 135hp engine. The bastard had my boat! At least Cummings had an eye for class. Man, was I jealous! I must crazy to have been thinking about a boat at a time like this!

Focus, focus. I had to focus on my predicament! I had to find an opportunity to escape! No doubt the Grand Banks would take us to our final destination and the link-up with Cummings's friend.

I was beginning to panic but quickly forced myself back to reality. My only hope continued to be Gary or someone from the team reacting when I didn't show up at the 3:00 a.m. meeting. The team would not have much to go on, but they were my only hope. I had to stall as long as possible.

My opportunity to break away from Cummings presented itself just before we reached the beginning of the dock.

Now!

I pushed Cummings in the back as hard as I could, spun left, and sprinted toward a pine forest using the van as a shield. I was praying that Cummings was honest when he said he could not kill me until I met his friend.

Two shots rang out, both close to my head but clearly not aimed to hit me. I zigzagged, keeping the van between me and Cummings. I was closing in on the edge of the forest.

Just a few more steps!

One more zig!

I heard the shot, tried to zag, but the third shot hit me in my left forearm, sending my head over my heels. I stumbled and finally sprawled onto the pine needles.

Son of a bitch, it stung like hell!

I grabbed my arm, stood up looking for another escape route.

Two steps and Cummings shouted, "One more step and I will kill you this time, asshole!"

Thank goodness I was right about Cummings having orders not to kill me, but I knew another break would be the end!

"You are one really dumb motherfucker, and luckily, I follow orders or you would be dead! Get his ass back to the boat, Jerry, and let's get out of here. It will be light soon, and we want to be in the ocean by then and make sure he doesn't bleed all over the place."

I staggered into the boat, holding my arm and trying to regain my breathing. My lungs hurt like hell, and I knew I had to stop the bleeding in my forearm. My already bloody shirtsleeve made a perfect tourniquet.

When Jerry was occupied, I managed to spread as much blood as I could in his goddamned classic boat! Not only would this piss Cummings off, but it is almost impossible to completely remove bloodstains, and maybe there would be enough evidence to help convict Cummings if I didn't make it through this!

It was important to keep conscious and alert. I couldn't give up. There had to be a way out of this mess. My marine survival training would guide me if I just kept my mind clear and pushed myself to respond the way I had been trained.

I would not let Cummings beat me!

Cummings told Randy to untie the boat and push us off and instructed Jerry to keep his gun on me. Cummings was at the helm. It was obvious he was familiar with operating his boat and quickly struck a course downriver, toward the ocean. He kept his speed reasonable, not wanting to arouse suspicion from anyone we might encounter on the river or the shoreline.

There was nothing I could do, so I used the time to sift through any possible event that could lead to an opportunity to escape. I needed to save my strength and stop as much of the bleeding as possible in my forearm. Another portion of my shirt made acceptable gauze to stuff into both sides of the bullet hole. "Mission complete," I said to no one in particular. "That should work."

There was no way Jerry was going to let me overpower him and take his gun. He would shoot first and suffer the consequences. That was not an option!

Diving over the side was also not an option, either!

Cummings would be on me before I could reach the shore, and the boat's propellers would make mincemeat out of me if he decided not to take me to his friend alive.

Anyway, I really wanted to know who this mysterious friend was. I was sure he was the person responsible for distributing drugs to the marines on the base.

We proceeded down the river for about, I estimated, ten miles and past the breakwater into the ocean.

There was not a cloud in the sky, like many summer Carolina mornings, and the reflection from the ocean hit me like a mirror pointed directly in my eyes, making it impossible to make out the direction we were headed, let alone spot a landmark.

Jerry was vigilant in keeping his gun on me. He gave me no opportunity to escape even if I thought I could.

I decided to screw with Cummings to pass the time! "Hey, Cummings," I shouted. He could hear me over the boat motors. "Do you have any idea where you are going, or are you just looking for a spot to throw me overboard?"

Cummings responded without turning around. "You'll see where we are going when we get there, and don't try to memorize the route. You won't be coming back!"

I asked again, "Do you really think you will get away with this? My guys will find me before you can do anything more to harm me. You might as well surrender now and make it easier on yourself! How about you, Jerry? Do you want to go to jail for a very long time for this guy?"

"Leave Jerry out of this, asshole!" Cummings said, before Jerry could respond. "Nobody is going to jail, and where you're going, only the fish will care! Now shut up and sit there like a good boy!"

I judged from the position of the sun that we had entered the ocean headed east.

Everything was in pain, but at least my nose and arm were no longer bleeding. I knew I had better take this time to recuperate as much as possible. I was not going to have too many chances to escape or defend myself.

I kept thinking the team would certainly be doing everything possible to locate me, but I also knew they would not have much to go on. I was sure no one saw Cummings dragging me to the van and drive away from the bar. Gary had left the bar about a half hour before Cummings arrived. He planned to go back to his hotel and change prior to the meeting with the team.

The boat was making exceptional headway in the calm morning ocean, running at what I estimated to be about fifteen knots. We had gone about eight miles when I recognized on the horizon where we were going. We were headed for the Tactical Aircrew Combat Training System (TACTS) ocean platform.

The TACTS was a redesigned fixed oil-drilling and production platform located ten miles off the South Carolina shore. The Marine Corps had purchased the platform and installed electronics in a one-story steel pole-type building to track pilot training over a forty-mile by eighty-mile grid. Sensors were placed in the ocean in patterns throughout the full grid. The data gathered by the sensors was transmitted back to the platform, calculated and forwarded to MAG-31 via microwave. The sensors tracked significant amounts of data, including each aircraft within the grid, and assimilated information on the coordinates of each plane, the relationship of each plane to others, airspeeds, altitudes, shifts in altitudes, as well as each gun lock achieved during a combat simulation. Everything was tracked and reported.

The platform was fixed in the sea with steel legs anchored on the bottom. Steel jackets made of tubular steel were piled into the seabed. Power was generated by solar panels and wind-powered generators, making the installation self-contained.

Cummings docked the boat next to one of the landing platforms and stairs leading to the deck, and motioned for Randy to lead the way to the top. Cummings followed and then me and finally Jerry, still holding his gun pointed directly at my back.

I made my way onto the platform and quickly surveyed my surroundings, looking for any possible avenue of escape.

Time was getting short!

There was a helicopter landing pad in an area to my right. The landing area as well as the rest of the platform was covered with wood floors coated with creosol for protection from the salt-filled ocean air. The railings around the landing pad and the rest of the platform were built with three horizontal bars approximately two feet apart and vertical stabilizing I-beams eight feet apart, also coated for protection. It would be difficult to scale the railing and dive into the ocean without bullets finishing the job!

My eyes were driven to a figure coming across one end of the platform. Obviously, this was the partner Cummings was so interested in me meeting.

Cummings waved and called to his partner, "There you are, Major. I wasn't sure you had arrived."

"Where's your boat?"

"On the opposite side of the platform."

"Is this the guy who fucked up our deal?" he said. He gave me an indignant once over.

"Morris, meet Major Todd Willingham, my partner," Cummings said. He pointed to Major Willingham.

"Yes, this prick is the fed. He has a lackey that my boys are taking care of as we speak!"

"My boys have cleared out the storage area in the bar and have taken the dope to my warehouse."

"Once this one is out of the way, we can make sure everything is okay there."

Willingham reviewed my appearance. He said, "Looks like he didn't want to cooperate! You know, I am really going to enjoy finishing what my friend here started," the major continued. "You have fucked up a beautiful operation and cost me millions of dollars. Now it's time you pay for that."

Willingham pushed me toward the rail on the east side of the platform. I was able to maintain my balance and swing my right hand, landing a direct blow to Willingham's head, sending him backward onto the platform floor.

Cummings grabbed me, smacked me across my mouth with his 9mm. "You shithead, you will never learn! I should shoot your ass right now and toss your useless body over the railing!"

"Stop it, Cummings," Willingham yelled, picking himself off the floor while rubbing the side of his head. Spittle flew as he placed his face directly in line with mine. "You think you're cute, don't you, motherfucker! Now, I am really going to enjoy each shot to your stupid head!"

Cummings said, "Jerry, Randy, get the ropes and concrete blocks over there," pointing to a corner of the helicopter landing area.

I finished wiping my mouth the best I could and, with a cold stare, said to the major,

"You're one sick son of a bitch. How can a marine do what you're doing to fellow marines? Obviously, the oath you took means nothing to you!"

"Fuck the oath," Willingham said. "I'm here to make money. I could give a shit what happens once I make a sale. I'm the guy who built the distribution on the base.

"You see," he continued, "my friend Cummings here gets the dope, delivers it to me, and I distribute it through guys I control on the base. Now you and your little buddy are trying to fuck that up."

I was trying to stall and sort out an escape plan. I said, "So, you get rid of me. Won't that kill your operation, and what about your dealers? How do you keep them from going to the MPs and spilling their guts for favorable treatment?" I asked.

"That was the sweetest part of the deal," he said. "No one knew where their shipments came from. All they got were cell phone calls with timing and directions. I was very careful not to expose myself to anybody. How could I and keep my identity safe otherwise!"

"Why bring me here?" I asked.

"I am responsible for this place. I'm the staff operations team leader on the base, so it's not out of the ordinary for me to be here, and nobody would think of looking here for you!"

"Enough of this bullshit," Cummings said. He was again pushing me toward the rail. "Jerry, get that shit over here so we can finish this guy and get on with it. I don't know why we had to come all the way out here to begin with."

"Jesus Christ, Cummings, how many times do I have to explain this to you!" Willingham said in exasperated voice. "We shoot his ass, tie him down good, and throw him overboard. The fish will take care of the evidence. We clear out and start up somewhere else in town!

"So, big boy," he said, pointing his 9mm at my head, "I can't truly say it . . . What the fuck is that?" The major turned and quickly recognized the marine UH-1 Huey helicopter approaching rapidly from the west.

Now was my opportunity!

I bolted across the platform and headed to the control building. I assumed the building was locked, so I dove around the corner just as Jerry unloaded his 9mm into the side of the building!

Randy decided not to be a statistic and headed for the east end stairs and Cummings's boat.

I made a run for him, diving with my arm stretched out as far as possible. I could use him as a shield if only I could get a grip on his work boot.

My fingers slipped across his boot, and before I could re-grip, he slipped out of my grasp, heading down the stairs.

I could see a Coast Guard MLB approaching at high speed.

They could handle Randy, so I turned to look for either the major or Cummings!

They were running toward the west end of the platform stairs and the major's boat.

Jerry wanted no part of the excitement. He threw his gun overboard and threw his hands into the air!

The chopper door was open, and a marine sharpshooter was covering both of them while another individual shouted through the chopper's sound system, "Drop your weapon and lay flat on the deck. Do it now!"

I recognized that voice; it was Gary. Thank God he was all right, but how did he get here?

Cummings fired several wild shots at the copter while running for the stairs.

The sharpshooter in the chopper opened fire, killing Cummings instantly!

I caught the major in a dead run by his legs, sending us skidding along the floor of the platform.

Now it was my turn! Two squarely landed rights and a solid left sent him curled up in a pile.

I had burns on my arm from skidding across the floor but really didn't give a shit! The major was on the way to a long posting in a military brig!

The chopper landed, and Gary jumped out, running to where I stood with my foot on the major. A marine was close behind and took control of Major Willingham.

"Thank God you are all right," he said. "I was afraid we would not get here on time."

"How in the hell did you know where I was? The last time I saw you, you were heading for the motel."

"I started there, but something told me I should hang around the bar until you left for the meeting. I've been covering your back during this whole operation, so I thought, why not this time!

"I saw Cummings and his gang arrive, but there was no time to warn you. I really didn't think Cummings would do anything at the bar, so I waited until he left with you in the van.

"I called Chief Knollson. He headed to the bar to arrest the guys who were moving the cocaine into the second van.

"I also called Agent Johnson, told him what was happening. I told him I would follow Cummings and give him our position as soon as I could.

"I recalled Agent Johnson when we reached the house and saw the boat. I requested he get there as soon as possible.

"Boy, I thought you were a goner when you made a break for it. I was afraid Cummings was going to kill you right there!"

"Cummings told me he was taking me to meet his partner, so I gambled he would kill me only as a last resort. I almost made it until the bastard shot me in the forearm!"

"Well, you scared the shit out of me! Anyway, after you left on the boat, Agent Johnson arrived, and we headed for the base and a helicopter.

"Major Anderson was the one who figured out where Cummings was headed. He guessed Cummings would not just dump you overboard and thought the TACTS platform was the destination."

"What would make Major Anderson think Cummings was headed for the TACTS platform?" I asked.

"Anderson had received information from Corporal Liston identifying Major Willingham's voice during the last telephone drop call, and Anderson knew he was responsible for the TACTS operation at the platform.

"Major Anderson decided to put a tail on him. Earlier today, Willingham launched his boat and headed for the platform wearing civilian clothes, so we knew he was not going there on military business.

"We put two and two together, loaded the team into the chopper, coordinated with the Coast Guard, and here we are!"

The marine team took Major Willingham below to the cutter to join Randy and Jerry and head for the base. The helicopter team swept the platform, loaded Cummings's body into the chopper, along with Gary and me, and headed home.

"Hot towel, sir?" The flight attendant's voice snapped me back to today. "We are beginning the breakfast service and will be landing in Shanghai in an hour."

I nudged Terri. She looked over at me, her hair disheveled from sleep. I handed her a hot towel. "Wake up, sleepy head, time for breakfast."

"Are we there already?" she replied. Terri was looking around the cabin.

"Already!" I said. "I don't know how you do it! I spent most of this flight reading and remembering Beaufort. We'll be landing in an hour, sweetheart. Hot or cold breakfast?"

"Cold. I haven't thought about Beaufort for years. Your friend Colonel Jackson has never forgotten that one or how you solved that case! If I remember right, Major Willingham was convicted and the DEA broke up the Charlotte ring and the Giordanos were convicted as well. What more could he ask of you!"

"You're right about that, and Chief Knollson was able to put two bad cops away for a long time, but I did almost get myself killed!"

"I know, honey, but you didn't!"

"I hope what lies ahead is not quite so dangerous!" I said. I looked at Terri. "I was a lot younger then! I hope I can do this. It's been such a long time since I have been involved in something like this."

Terri looked directly back at me and grabbed my arm. She said, "Come on, you may be a bit older, but you are still the best at solving these kind of puzzles. Little John needs us, and we can't let him down."

"I know, I know."

Our cold breakfast tray of cereal, banana for the cereal, a croissant with butter and jelly, yogurt, and a fruit tray arrived. Airlines always serve too much food in business class. We finished off our favorites and prepared to land in Shanghai.

Chapter 3

The flight landed on time at 10:30 p.m. Saturday night, May 10, and taxied for over a half hour to reach our arrival gate at international terminal A.

Shanghai Pudong International Airport was opened on October 1, 1999, replacing a smaller, older Shanghai Hongqiao International Airport that sat within the city of Shanghai. Pudong was the major hub in Asia and was located thirty kilometers from the city center with terminal A for international and terminal B primarily dedicated to interior travel. The airport currently occupied forty square kilometers of land adjacent to the coastline of the China Sea. Expansion plans would create an airport double in size with two additional runways by 2015.

Terri and I had been to China several times, and each time, my excitement grew as soon as I entered the terminal and proceeded toward customs.

The hallways leading to the customs hall and baggage claim were wide and brightly lit with gleaming marble floors. The hallways were decorated with Chinese art in display cases of various sizes and wall hangings depicting scenes from all sections of the Chinese countryside, each done in the artist's style. No matter how anxious or tired I was, the cornucopia along the way forced me to linger behind the rush to receive a passport stamp and official entry into the country.

The customs hall was virtually empty, and the customs agents were preparing to take a break before the next flight arrived from a destination somewhere in the world when Terri and I arrived.

We passed quickly through customs into baggage claim. Our luggage was sitting next to the posted arrival carousel looking lonely, as most luggage from our flight had been retrieved; however, ours sat waiting patiently to be recovered. We had nothing to declare, so we flowed with our luggage into arrivals beyond the secure customs and baggage area.

The arrivals terminal followed the same pattern as the hallways leading to customs, just more spectacular with soaring ceilings and huge curved windows beginning about twelve feet from the light beige marble floors. At night, lights adorning the V-shaped outdoor canopy sparkled like individual stars.

The ceiling light clusters of five individual lights were surrounded with elongated hanging simulated church organ pipes of stainless steel polished to a high resolution, giving off enough light to simulate daytime.

The late May evening weather was hot and humid when we exited to the area devoted to taxis. We had considered using the high-speed Shanghai Maglev Train, but decided the seven-minute train ride to the Pudong section of Shanghai was still not comfortable enough nor convenient enough to override the amount of luggage we carried away from baggage claim.

Two taxis later, one filled with luggage, we were on our way to the Holiday Inn downtown, located at 585 Heng Feng Road in Shanghai. The Holiday Inn was around the corner from Shanghai Railway Station, one of two that carried millions of passengers daily to all parts of China. I normally stayed in the Holiday Inn because most of my suppliers and Wright Manufacturing plants were located a reasonable high-speed train ride away. Our taxi ride from Pudong International took forty-five minutes, normal travel time to the hotel this time of night. Our driver moved expertly in and out of traffic, slowing to the speed limit only when he approached a roadside speed camera hanging menacingly over the highway on a silver gooseneck pole.

The Chinese love neon, and all along the expressway, primarily blue neon accented with either red or yellow lighting adorned most under—and overpasses, as well as bridges on our route to Shanghai. The American billboard has also invaded China. Brightly lit LED advertisements commanded most of the large roadside signs. The myriad of color the lights projected against the warm, clear night sky continually drew my eye to each oncoming sign as our taxi continued its journey into the city.

The Holiday Inn downtown consisted of two forty-floor towers with a large multi-story parking garage dividing them. The hotel occupied one full city block and was the dominant structure in the area. Office buildings surrounded the hotel complex with the bottom floors of each office building containing mishmash of food outlets, betting parlors, drug stores, restaurants, souvenir shops, and clothing outlets. Walking the streets around the hotel's twin towers felt like wandering into a minefield! There were sidewalk stands selling newspapers and magazines. Fruit vendors and carts were filled with Chinese vegetables and exotic herbs. Document couriers laid on their motor bikes, dressed in leather, waiting for an assignment or sitting on the curb talking and smoking cigarettes, complicating the process of moving along the sidewalk. The streets were filled with cigarette butts, portions of newspapers discarded after reading, drink bottles, and other assorted trash. Street cleaners in coolie hats with dirty, large, fan-shaped straw brooms fought a losing battle, sweeping against the wind created by the traffic on the street and sidewalk. All this with the sea of people constantly moving toward their destination made the streets memorable. The changing of a traffic signal caused a river of pedestrians to flood across the street, dodging bicycles and motorbikes who did not recognize a pedestrian's right of way.

There were more plush hotels in the area, but my frequent travels to China and my status as a Holiday Inn gold member plus the number of nights at the Holiday Inn in downtown Shanghai accorded me upgrades to suites on the executive floor. I requested my favorite one the thirtieth floor.

Chapter 4

I swiped the access card to room 3012 at 1:00 a.m. on Sunday morning, May 11. This suite overlooked the railway station and all the action of people in a twenty-four-hour Broadway-type spectacular. The suite was spacious, with a living room decorated with quality paintings of the countryside and containing a large-screen wall-mounted television, a small dining room with kitchen, and the ever popular minibar with a conference area for spreading out any information we might gather and to meet with those who would be assisting us to prove John's innocence.

A king-sized bed with the latest Holiday Inn bedding was the centerpiece of the bedroom, along with an oversized armoire containing clothes drawers and a second forty-two-inch television. Nightstands with ornate silver lamps with garnet shades sat on each side of the bed.

The two bathrooms were large and fully stocked with quality toiletries, telephones, and small televisions in each one.

Terri and I quickly unpacked just what we needed that night and climbed into bed. Now our roles were reversed. I was able to fall asleep quickly in a very comfortable and—based on my many trips to China and staying in this hotel—familiar bed. Terri tossed and turned for most of the night but did manage a few hours of sleep.

We decided the prior evening we would attempt on Sunday to meet with John at the police station where he was being held and to introduce ourselves to the high-ranking officials that would direct the investigation of the murder. We felt without the presence of a Shanghai counsel, this was a long shot but believed we had to start somewhere.

Our telephone calls to the Shanghai police were fruitless. No one was available to meet with us until Monday morning. We were instructed to call early in the morning for an appointment.

Bin Wang would join us on Monday for an early evening drink in the executive lounge at the hotel, followed by dinner. Bin knew several outstanding restaurants close to the hotel that were seldom frequented by tourists and were the equivalent of an American five-star restaurant. Why come to China to eat pedestrian food!

It was important to establish a baseline to begin the investigation, and we needed John's side of the story first; we would then meet with the Shanghai police to obtain as much information from them regarding what they officially knew about the case against John. I had done some research on the Chinese justice system and how the police investigation process worked during the two days it took us to arrange flights. It was clear from my research we would need support from a Shanghai law firm in order to access information from the police and to assist us in navigating the Chinese court system. We also needed to know more about how the police process worked. Bin had agreed to take the lead in obtaining a lawyer.

The Holiday Inn has four restaurants within the hotel, but the Western restaurant was used for breakfast. The executive lounge also served breakfast for those on the executive floors. I mostly chose to eat breakfast in the Western restaurant on the second floor. The room was bright and well decorated with Chinese art and sculptures. We arrived at 8:30 a.m. and were seated immediately at a table for four situated by the wall of windows overlooking the bustling city below. I normally chose the public restaurant over the executive lounge in order to mingle with the businesspeople and tourists from around the world who were staying at the hotel. I was continually surprised by how much I learned about what was happening around the city and near countryside regarding suppliers I might be considering or currently working with. Even though it was Sunday, the restaurant was busy. Almost all plants in China operated six days a week, and most businesspeople were in place on Sunday, ready to begin work early Monday morning.

A cornucopia of scents enveloped us when we entered the Western restaurant Sunday morning. The mix of Asian and Western food waffled across the room from several buffets that contained an enchanting variety of dishes. It was always a treat: mix dim sum, rice soup, fresh breads, fruit, and bacon and eggs for breakfast.

"My gosh, Rick, do you think you have enough food on that plate?" Terri said as I returned from my first round through the buffet. She could talk because her plate was barely covered with fresh fruit and one piece of Danish!

"I know, honey, but this is the one place I have a very difficult time making choices! It all looks so good, and I don't get this stuff at home!"

"Well, I understand, but I don't want you falling asleep when we are with John or with the police," she said. She was laughing.

I was ashamed to return to the buffet for seconds, even though I left some of my favorites behind, and I decided to start with them tomorrow!

During breakfast, Terri and I listed four initial questions we needed answered first. They were the following:

1. Why was John being held in a jail cell in Shanghai when the murder happened in the suppliers' plant in Changzhou?
2. Would John eventually be transferred there and which police force had jurisdiction and who would do the investigation?
3. What rights did we have in assisting John's lawyer during the murder investigation?
4. Did John actually have a law firm in Shanghai representing him?

We spent the rest of Sunday walking the streets around the hotel and resting. From prior experience, I knew whatever happened would not be to any plan we would derive, and we would need our strength to survive the emotional swings we were about to experience.

Chapter 5

We were standing in front of the hotel at 10:30 a.m. on Monday morning, May 12, waiting for the doorman to hail us a taxi to take us to the jail located at 128 Wuning South Road in the Jing'an District in the middle of Shanghai.

I truly loved Asia and never tired of the sights, sounds, and odors of the city streets. The countryside was completely different in appearance as well as smell. Even though it was mid-morning on Sunday, the city was alive with crowed streets, jammed with motorbikes and bicycles that maneuvered like a symphony written for motorized moving vehicles. Cars, though quickly becoming dominant, were forced to navigate with and against the tidal flow. There appeared to be no set traffic rules as there were in America. It was not unusual to see a car being driven on the wrong side of the road or turning against traffic or sitting in the middle of the road with the driver talking on his cell phone or looking at a map for directions!

Operators on two-wheeled vehicles were experts at placing their rides in spaces that defied gravity and the laws of physics; however, accidents were few, and it all just seemed to work!

Shanghai taxi drivers are the wonder of Asia! The hotel doorman translated our destination to the driver, and immediately upon the doorman closing the back door of the taxi, the driver bolted into the morning rush hour traffic without as much as a glance at traffic approaching in our lane behind us. Terri stifled a yip and stared at me with big green frightened eyes. I just shrugged and faced forward. It was too late to worry about what was coming behind us. I was focused on the traffic ahead and beside us. My intuition told me Nissan gathered up all their defective small sedans at one port and shipped them to Asia, specifically China, for conversion into taxis. Ours must have been an original. It was almost red in color with worn seats and seat covers; sitting in the backseat was for short rides only. The driver paid no attention to the rattles and engine noises, so I assumed they were normal. He darted in and out of traffic, strategically placing his cab in position to be first off upon the traffic signal change. He never appeared to notice the plethora of bicycles competing for the same space on the road and miraculously delivered us to the Wuning South Road Shanghai Municipal Police Station in less than thirty minutes.

The Shanghai Municipal Police Station occupied a noteworthy portion of one city block in the center of an upscale residential area where high-rise apartments displayed modern examples of Chinese architecture, and major executive Asian and American chain hotels stood as monuments to the phenomenal growth and Westernization of Shanghai and China. The only indication that the imposing, though plain, sand-colored building housed a key command center of the Shanghai police was the continual flow of uniformed police officers and average citizens in and out of the undistinguished front doors of the station. There were no police cars visible in the front of the building, and the sign on the right side front wall of the building identifying the station was small, ornate, and circular, predominately gold, featuring the logo of the Shanghai Municipal Police.

Our taxi driver understood enough English to thank us for the tip and proudly point out the station for us. We thanked him, and I said the blessing I always say after a Shanghai taxi thanking God for my surviving another perilous ride!

A huge ten-foot-high semicircular counter manned with at least ten officers was directly in front of us when we entered through the front door. There was a standing rail for exchanging documents between those visiting the station and the constables manning the desk. We could see an enormous area completely filled with desks, chairs, filing cabinets, and glass-partitioned offices behind the counter, all filled with various types of people talking to a constable each with a particular problem that demanded attention.

Typical for China, there was no organized queue of people entering and waiting to be called to the counter to state their problems. It was every person for him—or herself as people pushed and shoved to be the next person to reach the counter. Many years of travel and working in Asia did not make me feel rude as I pulled Terri behind me and made my way to the front. We approached what appeared to be a sergeant sitting in the middle of the counter when motioned forward. His uniform and attitude made it clear he was the gatekeeper.

"*Ni hao*," I said. I did my best to make my voice very respectful and my Chinese understandable. "*Ni hui shou Yingyu ma?*"

The sergeant glared down at me and said with an official voice to match my respectful one, "Ni hao, yes, I do. What can I do for you and the person with you?"

"Sir, my name is Rick Watson, and this is my wife, Terri. I understand that John Alworth, another American, is being held in this station. We have travelled from the United States to see him, if possible."

The sergeant turned to a constable sitting next to him and spoke quickly and firmly in Chinese, apparently giving orders to him on what to do with us. After giving his orders, the sergeant turned back forward, ignored us, and motioned for the next person to come forward.

The constable left his black swivel chair and proceeded to another counter behind him, obviously containing forms and various other information required to advance beyond the check-in counter. "Fill out form and give back to me," the constable said. He handed me a form and pointed to a waiting area in an adjoining room.

"*Xie xie*," I said. I took the form as we headed that direction when we were interrupted by another voice. A voice far from being official but portraying authority!

"Mr. Watson, may I be of service? I am Feng Shou, chief superintendent of police." He was large for Chinese standards and wore glasses that constantly fell down the bridge of his rather protruding nose. His complexion was darker than most Chinese and weather beaten. It was obvious the chief superintendent had spent a substantial portion of his career working outside in the elements.

"Yes, you may, Chief Superintendent," I said. I did not use any Chinese phrases because clear English was being spoken. "My wife and I are here regarding the murder of Zhu Zhong Huang, who worked for Universal China Production Company in Changzhou. I understand Mr. John Alworth is being held in this jail for the murder. Mr. Alworth is a friend of ours, and he requested we come to China and do what we can to help prove his innocence."

The chief superintendent drifted his eyes from Terri to me like a film director absorbing the set and how the actors were positioned prior to his yelling "action" and filming the next scene in a movie.

"Please, follow me to my office so we can discuss." We followed the chief superintendent to a private elevator. The elevator, located at the back of the first floor of the station, was typical for an Asian elevator: very small and slow! Nothing was said during the ascent to the chief superintendent's office, located on the second floor.

The second floor was the antithesis of the first floor. The entire floor, however, was organized with private offices and small labs. I felt as if we had just stumbled onto a set from any of the *CSI* television crime shows. Terri looked at me, and I immediately knew she was thinking the same thing. "I wonder if the Chinese watch *CSI: Miami!*" I said quietly to Terri. She laughed a little too loudly, causing the chief superintendent to glare at both of us.

His office was in the middle of the building along the west wall. It was spacious but sparsely decorated, with a few personal awards and letters of promotion and accommodation. The office looked directly into an apartment building next door. The furniture was serviceable but dated, with little style. It consisted of a desk with two chairs facing it and a small conference table with six chairs all made of an unrecognizable wood. The furniture appeared to be miniature, based on the size of the office.

Government buildings were one of the last outposts where the Chinese Communist Party was displayed prominently. The chief superintendent's office would have won a first-prize ribbon for party displays had there been such a contest. There were pictures of the current premier, Wen Jiabao, posters of Chairman Mao, party banners, and several pictures of the chief superintendent with other police and party officials.

He pointed to the two chairs positioned in front of his desk, requesting that we sit while he assumed his place behind his desk. "So, you are friends of Mr. Alworth," the chief superintendent said, looking out the window, not directly at Terri or me. "What makes you think you can help him? You have no authority here. Besides, we are very capable of handling cases such as these," he continued, not allowing either Terri or I the opportunity to respond. "Murders are happening all too frequently in China now. Our young people are contaminated by Western television, movies, and music full of violence! My country unfortunately is heavily influenced by the money your country now spends on goods to be exported back to you. Many people have made fortunes working for you, but it has only lead to no good!" I was surprised at the hostility toward the United States displayed by the chief superintendent. Most of the Chinese businesspeople

I knew truly liked Americans and were progressive in their approach to the continued development of the country.

He turned in his chair, looking directly at us with almost a look of contempt. "I spent many years in the United States in university and working with American police forces in New York and Los Angles. I learned many things about the treatment of criminals from your justice system. We are not as lenient as your country on criminals. Your Mr. Alworth should have been more cautious and thoughtful before taking the life of a citizen of my country. We believe punishment should match the crime swiftly and completely!"

"Chief Superintendent Feng, over the past twenty years, I have travelled throughout China on a frequent basis for business and pleasure, and have seen your fine police departments in action. I would agree your country is different than the United States in controlling crime, including murder. I, too, am concerned with our justice system at times; however, our friend, Mr. Alworth, has also been in China many times as well and recognizes how crimes and criminals are dealt with here. Mr. Alworth is a cautious man and has great respect for the laws of your country. That is just one of many reasons why I believe Mr. Alworth could not have been involved in any murder!"

"I see; however, I am very sorry, but you see, it is impossible for me to help you today. It is true your friend is here in the building, but unfortunately, he is not available for visitors. I am aware there has been a murder committed, but I cannot tell you more at this time. Now, if you will allow me, I will show you to the building exit and arrange for transportation back to your hotel," he said. He unceremoniously, clearly ended our meeting.

It was obvious we were going to get nothing from the chief superintendent without legal assistance from a Chinese law firm, so we thanked him for taking his valuable time to see us and sharing what information he could at this time. We followed him to the elevator that would take us to the first floor. The chief superintendent was silent on our decent and personally escorted us to the front door.

A black Buick Century was waiting for us at the curb as we exited the building with the driver in place. The chief superintendent gave him the address of our hotel and wished us a pleasant stay in China as he closed the door to the backseat of the Buick. He stood beside the car and waved goodbye as we pulled away from the building into city traffic. Terri and I talked of immaterial things during the ride back to the hotel, knowing the driver probably understood English and was under orders to report our conversation back to the chief superintendent.

We arrived back at the Holiday Inn at 1:30 in the afternoon and made our way to the executive lounge to sort out what happened at the police station and our meeting with the chief superintendent.

Two beautiful uniformed young women seated at a reception desk greeted us as we entered the lounge and asked for our room number. One of them left the desk to lead us to a booth along the far west wall of the lounge, directly across from the doorway to the back room where food received from the kitchen and drinks were stored.

A simple buffet of fruit, nuts, breads with cheese, and Chinese cookies was laid out in the center of the room. A fully stocked bar also with soft drinks was on the wall opposite the buffet, along with a coffee machine and hot water for various types of Chinese tea.

The attendant politely asked our pleasure, slightly bowing as she asked. Terri requested fruit juice, and I asked for a cup of my favorite Chinese tea. We both realized we had not eaten since breakfast, so we helped ourselves to the buffet after our drinks arrived.

"Well, that trip got us nowhere! I didn't expect much, but I thought we would at least be allowed to see John." Terri began the conversation after some fruit and cheese.

"I was surprised at the attitude of the chief superintendent," I said, "but we said before we left Charleston this wasn't going to be easy. Looks like the first thing we have to do is locate a reputable Chinese law firm familiar with cases like John's to represent him and assist us. I'm sure Bin can help us find the right firm. He has been here long enough to know who the real players are. We will make him earn his dinner tonight!"

Terri said, "I know one thing: I'm ready for a power nap before we meet Bin. Jet lag is catching up with me!"

We finished our drinks and our light snack, and then headed for our rooms. I told Terri I needed to walk a bit to think about what happened with the chief superintendent and get mentally prepared to talk with Bin, so I made sure she was safely in our room and took the outside wall elevator down to the lobby.

I let my mind wander as I watched the street below slowly go from miniature to life sized as the elevator made its way to the first floor.

It took only a few seconds to meld into the sea of black-haired Chinese people determinedly swimming to their destinations. I walked across the street and found a seat on a bench facing the enormous big screen monitor that is the centerpiece of the train station plaza. Some people need to be alone when they think; I always preferred to be in the center of the action and noise level. I have always been able to block out the sound and absorb the events transpiring around me.

For the next hour, I mentally prepared the story, as I knew it, for Bin and my thoughts on steps we should be prepared to take to prove John's innocence. I needed to know more about Gregory Brightson and anyone else involved with John in the development of the product and what John learned from the head engineer at Universal China Production Company prior to his death. What happened to the information Zhu was going to give John, and what did it say? What could be so dangerous in the product that could cause someone who used it to potentially die? It had to be in the hardware; nothing in the software application could cause any failure like the one John described to me. The software only controlled the application.

Who did John think could be responsible for killing Zhu Zhong Huang, and why would someone want him dead? I needed time to visit the plant, look at the production floor and the product being produced and any engineering and quality data I could get my hands on. There has to be a smoking gun somewhere in the plant and operations data.

Who had the most to lose if the product failed? Brightson, the owner of Universal China Production Company, or someone within or associated with WMT.

Could the killing be completely unrelated to anything John was working on? Maybe bad blood between Zhu Zhong Huang and someone inside Universal China Production Company? Not likely, but we needed to close all the possible ideas and leads as quickly as possible. I had a feeling the Shanghai police or the Changzhou police would want to solve this case as soon as possible to make an example of it to the local Chinese and the American people who work in China. Where would the case be investigated? John was taken from his hotel room in Shanghai, but the murder did happen in the plant in Changzhou.

Before I knew it, rain began to fall, and I hustled back to the hotel, just barely beating a Shanghai downpour!

Terri was awake and reading when I entered the suite about 4:30 p.m. "Looks like you still can put it in high gear when you need to! I watched you running across the plaza!" she said as I used a bathroom towel to dry off as best I could. "Ann called when you were out thinking. She is on her way over to see us. I asked her to call us when she arrived and said we would meet her in the executive lounge."

"Why not here?" I asked.

"I just thought she would be more comfortable in the lounge. You know how she is about status, and the executive lounge does represent status!"

About twenty minutes later, Ann called. Terri and I took the elevator to the executive lounge. The elevator door opened on our floor, and Ann was standing in the back looking like her splendid self. I was convinced raindrops wouldn't dare to fall on her!

"Terri!" she exclaimed. She grabbed her with as tight a bear hug as she could. She began crying as she held on to Terri, "My God, I am so glad to see you both. You can't believe what it has been like these last few days." Releasing Terri, she gave me the same type of hug. "You just have to get my husband out of this. You know John could never do something like this. He could never kill anybody!"

The elevator door opened, and we proceeded to the lounge. I signed Ann into the log book when requested by the same young woman who was manning the lounge earlier in the day. We sat at the same booth we used for drinks and snack, Ann and Terri on one side and me on the other.

Drinks arrived, and Terri asked Ann, "How are you holding up?"

"I'm doing as well as can be expected for someone whose husband has been falsely accused of murder in a foreign country! It will be better now that you both are here!"

"Ann, do you have any idea who could be responsible for the murder of Mr. Zhu?" I asked.

"I have no earthly idea, "she replied. "John always got along well with everyone at UCPC. He had no problems with anyone there that I know of," Ann replied.

Terri asked, "Have you talked with John?"

Ann took a deep breath and began, "Those assholes at the police station won't even let me see John, even though I am his wife. They just keep saying he is being confined to solitary until a decision is made on how to proceed with the case. I could just kill the bastards!"

"Easy, girl," I said. "One killing is enough for now!"

Terri asked, "What has WMT done? Have they done anything for you?"

Ann managed a small laugh and continued, "I have talked to both Gregory Brightson, John's boss, and Dennis Dureno. Gregory will be here tomorrow, and Dennis is already here."

"Who is Dennis Dureno?" I asked.

"Dennis is the vice president of marketing for WMT. He arrived here a week before John."

"Why is he here?"

Ann replied, "I'm not sure. John said something about how he was going to meet with the government and some potential Chinese customers who might be interested in the new product for the domestic market."

"I assume he is aware of John's situation?"

"I'm not sure. He is staying at the same hotel John and I always stay at. He had breakfast with John and me just prior to his leaving on his current trip inside China."

"Was he aware of the problem with the product?" I asked.

"No, I don't think so. If he was, he didn't share it with me," Ann replied.

"Is he still somewhere in China?"

Ann replied, "Yes, I believe so. I haven't seen him at the hotel."

I said, "I would like to talk to him when he returns."

"Why hasn't the company retained counsel?" Terri asked. She put her arm around Ann's shoulders for moral support.

"Gregory told me he would take care of finding the right firm when he arrived. For me not to worry, he would make sure John was taken care of."

"What time will Brightson arrive?" I asked.

"I think at ten thirty p.m. tonight. He is also staying at the Shangri-La. I know how desperate Gregory is to have his product go into production, so I sure hope he really means it when he says he will help John. I am afraid he will do just enough to show others he tried, but let John be found guilty!" Ann said. She sobbed into Terri's shoulder. "That's why John needs you on his side to find the real killer."

"Well, we will do what it takes to prove his innocence," I replied. I looked at Terri. "I have good friends here who will support and work with us. One of those people is Bin Wang. He will be here within the hour to join us for dinner. Would you like to meet him and join us?"

Ann replied, "I'm sorry, I can't. I have a scheduled dinner with a new client that I can't miss. Who is Bin Wang?"

"Bin is an old colleague of mine who runs his own business here and has excellent contacts."

"Oh, I see," Ann replied.

"Honey," Terri said, "don't you worry; we'll find who really killed Mr. Zhu."

"I'm sorry, you guys, but I really must go, "Ann said. "I don't know how I will ever be able to thank you enough for your help. Both of you are our best friends, and we love you. What are your plans from here?"

I replied, "I plan to select a law firm to represent John and to get into the jail to see him tomorrow."

We walked Ann to the elevator outside the executive lounge, and after goodbye hugs and reconfirming our commitment to free John, we returned to the lounge to discuss the events of the meeting with Ann and our plan of discussion with Bin.

Terri started the conversation. "Rick, did Ann seem a little distracted to you or not as concerned as I thought she would be?"

"You know how Ann is; it's hard for her to always show her emotions, and her outside is always harder than her inside," I replied. "However, I thought it odd she chose meeting a client over having dinner with us and meeting Bin."

"I guess you're right, but it still seems odd to me. What's next, honey?"

"One, I want to talk to both Dureno and of course Brightson. Second, we need Bin to help us find a good law firm familiar with this type of case either here or in Changzhou. Third, Bin needs to pick a good restaurant for tonight. I'm starving."

"Okay, okay! I need to freshen up before he gets here and change into something more suited for dinner," Terri said.

"Yeah, me, too!"

"Oh please! Let's get out of here," she said.

Bin called from the lobby right on time at 6:30 p.m., just as Terri finished getting beautiful, and I told him we would meet him in the executive lounge in a few minutes.

We arrived in the lounge and I was signing the guest log when Bin arrived. "Ni hao, Mr. Bin," I said. I gave Bin a hearty handshake. "It's great to see you."

Bin was about five feet five with a slight build. He wore glasses and had short black hair like most Chinese people, and there was always a smile across his broad, handsome face. He had met his wife Holly in university, and the marriage had produced one boy, Charles, who was now twelve.

Bin gave Terri as close to a hug as you will encounter in China and said, "You both look just as wonderful as the last time we were together."

"Oh, hell, Bin, we just saw you six months ago when Terri and I were here on a shopping holiday! Certainly we couldn't have changed that much!" I said.

"How could I forget that trip!" Bin said. He winked at Terri, who loved to shop in Shanghai and dragged Bin along with us for support. Bin, the ever consummate polite Chinese person, always smiled and followed her lead. It was a good thing he loved Terri!

"Let's sit in our newly favorite booth so we can discuss in private why we are here," Terri said, leading us to the booth in the back of the lounge.

After receiving our first round of drinks, I began telling Bin the story that precipitated our trip.

I completed my story after answering all of Bin's questions along the way and asked Bin if he could help us locate law firms in both Shanghai and Changzhou to represent John. It was obvious we would need one firm to assist us in gaining access to John in jail in Shanghai and one firm to represent him if the case and investigation were transferred to Changzhou.

"No problem," Bin said. "I will get that done by midday tomorrow. I know you don't believe John had anything to do with the murder, but do you have any clues to who may be responsible?"

"Not yet," I said. "We were not allowed to see him when we visited the Shanghai Municipal Police Station on Wuning South Road. We met with John's wife, Ann, earlier today, but she had no idea who would have wanted Mr. Zhu dead."

I continued, "It's seven thirty, and I've been hungry since this afternoon. Bin, I'll buy if you pick one of those excellent Shanghai restaurants you are famous for!"

"No problem, I know just the right place. If I remember, Terri raved about the food the last time you were here. Let me call right now and make a reservation for us."

"I certainly hope you are calling Baoluo! Do you really think you can get us in with such short notice?" Terri said hopefully. Bin introduced us to

Baoluo on our last trip to Shanghai and Terri rated the restaurant right up there with Bistro Savannah and that was an accomplishment!

"I know the owner fairly well, so I hope he will make room for us," Bin replied as he dialed his cell phone.

We were in luck and were awarded a reservation for 8:30 p.m. Perfect timing!

There is an old saying in China: "Food is the first necessity of the people." Food is a part of Chinese culture and is socially treated as a communal affair. It is most often taken together as a family unit with emphasis on shared bowls presented normally on a lazy Susan. The most senior or elderly person or guest is seated at the center of the table and always begins the meal by taking the first bite. Legend has it Chinese cooking originated with Yi Yin, a minister in the Shang Dynasty (ca. fifteenth to eleventh century BCE). Chinese cuisine can generally be classified under southern and northern categories. Over time, many distinct local flavors were developed and grew into categories of their own.

Baoluo was famous for incorporating several of these categories into their menu. The restaurant located at 271 Fumin Lu appears from the street side to be a very tiny diner occupying one unit on a street of tightly packed Chinese houses; however, once inside, you see the restaurant actually stretches four houses deep. The story told over and over says the original owner ran a bicycle shop and lived on the lane. He started the restaurant business in his home and purchased his neighbors' homes as his business expanded.

Terri's favorite dish, *Huiguo rou Jialing*, twice-cooked lamb wrapped in mini pancakes, was also a favorite of mine, but I could not pass up the fatty, bad for you *hong shaorou*, braised pork belly. Bin always insisted that *songshu hyu*, sweet and sour fried fish, also be a part of the meal. Our discussion during the cab ride to Baoluo was dominated by the number of dishes we would order upon arriving at the restaurant and laughing at the Chinese tradition of watermelon always signaling the end of the meal!

After dinner, we began discussing John and his case. "Bin," I asked, "Can you help us find a good law firm here in Shanghai as well as Changzhou? We need a firm with a solid reputation of dealing with murder cases."

"No problem, I have friends in both cities who can help us. I will make arrangements for you and Terri to meet with a Shanghai firm tomorrow morning," Bin said. "You know, when I was younger, there were no firms in

Shanghai who handled murder cases. There weren't enough murders each year to generate enough business. It's funny how time changes things!"

We finished our exquisite dinner about 10:30 p.m. Bin took a cab home, and Terri and I returned to the hotel. Bin promised to call mid-morning on Tuesday with our appointment time with the Shanghai counsel.

Chapter 6

Terri and I had just returned to our room around 8:30 a.m. on Tuesday, May 13, after breakfast in the executive lounge. She nor I was up for a large buffet breakfast after our excellent dinner the previous evening at Baoluo. The ringing room telephone was Bin. He had contacted Lu, Wxi, and Hoa and arranged a meeting with Ms. Lily Chen, one of their lawyers, at 11:00 a.m. that morning. He said he would meet us in the lobby at 10:00 a.m. to brief us on both the firm and Lily prior to our meeting.

We were drinking Chinese tea when Bin arrived. He joined us and began his brief. "Harry Wxi is an old friend and heads one of the top criminal law firms in Shanghai, Lu, Wxi, and Hoa. I called him this morning and he suggested we meet with Ms. Lily Chen after I described your friend's situation. Ms. Chen is a partner in the firm and a well-known criminal defense lawyer. I spoke to her of your visit to Shanghai Municipal Police Station and your meeting with Feng Shou, chief superintendent of police. She was not surprised at the outcome and suggested she listen to your story and then decide if she would be in a position to help you."

We thanked Bin for acting so quickly to help us and asked if he would accompany us to meet with Ms. Chen. "No problem," he replied.

The taxi ride to 1601 Nanjing Road West took thirty minutes. We arrived in the twentieth floor lobby of Lu, Wxi, and Hoa on time at 11:00 a.m. Achievement radiated throughout the vast, vaulted ceiling lobby decorated in muted brown and dark tan colors containing the most tantalizing collection of Chinese art and sculpture I had ever seen outside the museum of art in Shanghai. I could have spent the entire day in the lobby just absorbing the displays surrounding us. A striking hand-carved oversized antique desk that matched the other furniture pieces impeccably dominated the reception area. We had just settled into living room—comfortable black easy chairs when Ms. Chen's assistant welcomed us and led us to Ms. Chen's office on the twenty-first floor. Even the elevator was a work of art!

Ms. Chen's spacious office, befitting a company partner, was strategically positioned on the east corner of the twenty-first floor with windows that rose from deep pile walnut-colored carpet to the ceiling. The view was stunning. On a clear day, a substantial portion of Shanghai was visible. If office position meant anything, Ms. Chen was extremely powerful within the firm! Her office was a continuation of the décor in the lobby with oddly very few personal affects anywhere within view.

"Mr. Wang, it is a pleasure to see you again so soon," she said. She extended her hand to Bin, crossing from behind her desk to the entrance of her office.

"My pleasure, Ms. Chen," Bin replied.

Ms. Chen was attractive and very small, with black hair styled to accentuate her face impeccably. She wore a designer green tailored business suit that matched the fashion prevalent in today's higher level business circles in China. Her jewelry appeared expensive, creating the desired accent. She turned, facing both Terri and I. "I was with Mr. Wang and his lovely wife, Holly, last week at the annual charity ball for children with cancer."

"Ms. Chen, may I introduce Mr. Richard Watson and Mrs. Terri Watson, his wife," Bin said.

"It's a pleasure to meet you both," she said. She shook our hands with a firm but delicate grip. "Mr. Wang has been kind enough to provide me with some details about your friend's predicament. I truly hope I may be of assistance."

"Thank you so much for agreeing to see us on such short notice. I know you must have a full schedule of cases that need your attention. Also, please call me Rick. 'Mr. Watson' is too formal!"

"Thank you. Now, please." She pointed to a long couch and loveseat positioned to take advantage of the spectacular scenery outside her office windows. "Tell me about your friend."

"Terri and I have known John Alworth for over twenty years, and he has never displayed the type of personality that would lead anyone to believe he would be involved in a murder. He is currently director of quality for Water Management Technologies, a rapidly growing startup company in the United States who developed new technology for managing water consumption and energy in the home. John selected Universal China

Production Company, located in Changzhou, as the supplier to produce the hardware and integrate the software developed in the United States by Water Management Technologies into the final product.

"WMT is planning to go public after a successful launch of the product into the United States consumer market. John arrived in Changzhou on Monday, May 5, to meet with the owner, Da Wei Huang, and Zhu Zhong Huang, chief engineer of UCPC. He was there to give final approval and authorize production of the new product.

"All we know is John held the meetings as scheduled. During his meeting with the chief engineer, Zhu Zhong Huang, Zhu informed John there was a drastic problem with the product and the final testing. According to John, the product developed a flaw that Zhu said could be fatal to users if not corrected. Zhu would not describe the failure to John in the plant but agreed to meet him at John's hotel later that night with the data. John thought that was strange, but agreed. All John knows is that he was tired from jet lag and inadvertently fell asleep prior to Zhu's arrival. Next thing John knew, the police were at his door and arrested him for the murder of Zhu Zhong Huang!

"He was taken to Shanghai Municipal Police Station, where he used his one telephone call to call us in Charleston five days ago to come help him."

"Why did Mr. Alworth call you instead of his company? They would appear to have been the logical first call for help."

"John told me when he called that Terri and I were the only people he could trust. He was afraid to call Gregory Brightson, CEO and chairman of WMT, because Gregory was determined to push WMT public on schedule, and any change in the production start would cause a miss of the critical date. John was afraid Gregory might not believe him."

Ms. Chen said, "When did you go to the police station?"

"Terri and I were there yesterday morning and met with Chief Superintendent Feng Shou. He was not cooperative and politely informed us there was nothing we could do."

"In fact, he was just downright rude regarding his opinion of the United States," Terri said.

"I believe he does harbor considerable resentment of your country," Ms. Chen replied. "However, I can circumvent that with the proper objections and demands to see my client."

"Does that mean you are willing to assist us?" I asked. I was surprised at how quickly Ms. Chen agreed.

"Yes, I will take the case. Mr. Wang is an old friend of the firm, and I am intrigued with what I have heard so far. Now, please tell me, who are the main players in this case you have not mentioned as of now?"

Terri took over the conversation. "First is Ann Alworth, John's wife. She is here most of the time with John and has started a business assisting new, mostly American, families who have been assigned to China for the first time. Second is Mr. Gregory Brightson, CEO and chairman of WMT. Rick mentioned him earlier. Next is Mr. Dennis Dureno, vice president of marketing for WMT. Neither Rick nor I know him, but he is a frequent visitor here. Mr. Brightson will arrive here tonight, and Mr. Dureno is already here, but somewhere in the countryside."

"I mentioned both Mr. Da, owner of UCPC earlier, and of course Mr. Zhu," I said.

Ms. Chen replied, "So, you know of no one else at this time?"

"No, that's the list," Terri said.

Almost as an afterthought, though I knew it wasn't, Ms. Chen asked, "Mr. Watson, do you know anything about the product or what could cause the catastrophe Mr. Alworth believes he heard from Mr. Zhu?"

"No, unfortunately, I have not been involved with John in a business relationship for some time. All I know is what he has mentioned in passing prior to our trip regarding the incredible opportunity this product would provide everyone who worked for WMT."

She continued this line of questioning. "What about Mrs. Alworth? Would she have any knowledge of the product? Sometimes wives are deeply involved in their husbands' work."

"No," I replied, "Ann Alworth was not the type to want to be involved. She would be more interested in the opportunity!" I said. I was afraid my answer may have painted Ann in a bad light.

"Well, thank you, Mr. Watson, Mrs. Watson," Ms. Chen said. She nodded toward each of us.

"I will organize the necessary paperwork to introduce myself as Mr. Alworth's legal counsel and request a time from the Shanghai police to meet with him. I will include both of you, if possible. There might be a problem with that request, but we will make it anyway. I will have the paperwork delivered to Shanghai Municipal Police Station this afternoon and request a visitation for early tomorrow morning," Ms. Chen said.

"If the decision is made to prosecute in Shanghai, I will need to interview everyone on your current list and work with you to develop a working strategy. If not, I will pass my information on to whomever you choose to work with in Changzhou. Our firm has working relationships with the best firms there and can make a recommendation at that time, if you wish! For now, let me get to work. I will call you when I have something to report."

We rose to leave her office. Ms. Chen asked, "How many times have you been in China, Mr. Watson, and how did you and Mr. Wang meet?"

"I have been here thirty-two times, counting this trip, beginning in the early eighties. And I have been here seven times over the past several years," Rick said.

"I trust you have seen immeasurable changes since your first trip, Mr. Watson. I hope all for the best," Ms. Chen said. A look of amazement crossed her face!

"No doubt. I have grown to love China and always look forward to my next visit. It has been a privilege to have travelled through most of this great country."

Laughing, Ms. Chen replied, "You have probably seen more of China than I have! And what about Mr. Wang here?"

It was Bin's turn to comment. "Mr. Watson and I worked together for many years prior to Mr. Watson's retirement. First, at Wright Manufacturing, when I was director of supply chain for Asia, and occasionally now that I own my own supply chain company."

"Very interesting, I believe you two would have been formidable opponents for our Chinese suppliers to your company!" Ms. Chen said. She walked toward the door of her office.

We said our goodbyes, thanking Ms. Chen profusely, and followed Ms. Chen's assistant to the elevator that would take us to the lobby. Ms. Chen's assistant held the elevator open while Bin said his personal goodbyes. We made our way to the elevator that would take us to street level after arriving back in the lobby. I took one lingering final look at the exquisite lobby décor before the doors to the street elevator closed.

Once outside the building, we agreed to call Bin when we heard from Ms. Chen and thanked him for taking time to help us.

"No problem, however, now I must return to my office and resume my daily schedule," Bin said. "Please do call when you have more information."

We both hailed taxis and headed our separate ways, us back to the hotel and Bin to his office, after saying goodbye to each other. Terri received a Chinese hug, and I received a handshake!

It was after 1:00 in the afternoon when we entered the taxi, and we realized we were hungry. It had been hours since we had enjoyed our small breakfast. We requested the driver take us to the Golden Dragon Restaurant located on the Pudong side of Huangpu River that splits old and new Shanghai. The Golden Dragon was near the Orient Pearl TV Tower and had been a favorite of ours for years.

The Golden Dragon was normally busy no matter when you arrived, and today was no exception. A prime location with excellent food! We waited about twenty minutes for a table. Luckily, our table was window side, overlooking the river.

The Huangpu River is busy365 days a year, twenty-four hours per day, moving a wide variety of consumer and production goods on large, flat river barges, small boats piloted by heads of families who live and make their living aboard their boat, tourist sightseeing boats and dinner boats docked at the Bund, advertisement boats, all sizes of tug boats, and even small cruise ships that docked on the old Shanghai side of the river, a short stroll from the Bund. Occasionally, a small Navy boat would make its way down the river as well. Today's air was less smoggy than what had become normal with the unchecked industrial development in the Shanghai area, so we thoroughly enjoyed the river dance playing before our eyes as we devoured our lunch.

"I don't know about you, but I'm ready for a swim in the pool and a sauna," Terri said. She finished the last of her tea. "This has already been a long trip, and we have only been here two days!"

"Sounds good to me. I think I'm ready, too. We missed our daily run this morning. In fact, how about dinner tonight in the Great Wall Tower?" I said.

"Sounds delightful," she said. She knew tonight's dinner would be no match for our night at Baoluo! "You pay while I freshen up, and then on to the hotel!"

Both the swim and sauna at the hotel were as excellent as they sounded in the restaurant, and dinner was more than acceptable in Angels Cafe on the second floor of the Great Wall Tower.

Chapter 7

Terri and I were up early on Wednesday, May 14, as our acclimation to the twelve-hour time change took hold. At 7:30 a.m., we were running on side-by-side treadmills on the thirty-sixtieth floor exercise club. Our normal daily routine began with a stretching program, an eight-mile run (normally outdoors), strength and flexibility conditioning, followed by more stretching. For me, a small amount of stretching and the run would suffice, but Terri insisted the full boat would keep me young, so I learned to not resist long ago and followed her lead. Today, we also included a trip to the sauna.

We had just returned to our room at 10:00 a.m. when the telephone range. It was Ms. Chen's assistant, requesting we meet with Ms. Chen at 1:00 p.m. in her office. We quickly agreed and called Bin to inform him of the meeting time. Bin apologized, saying he had a scheduled meeting with a major customer that afternoon and requested we call him after our meeting with Ms. Chen.

Ms. Chen greeted us warmly as we entered her office promptly at 1:00 p.m. If a second meeting was any indication, her daily pattern of dress was a business suit. This time, the suit was black, with shoes and jewelry carefully selected to accentuate the outfit perfectly.

"Good afternoon, Mr. and Mrs. Watson, thank you for coming. I have reached an agreement with the Shanghai police regarding Mr. Alworth, my representing him, and your involvement, Mr. Watson."

We were seated at the conference table in Ms. Chen's office when she began, "I am meeting with Chief Superintendent Shou today at 3:30 p.m. to discuss why Mr. Alworth is being held without charges and what evidence the municipal police have that incriminates him. I will demand the police either charge Mr. Alworth or release him on bond within twenty-four hours. Unfortunately, you will not be allowed to attend this meeting; however, I have an agreement that you, Mr. Watson, will be allowed to see Mr. Alworth no later than tomorrow, Tuesday.

"For now," Ms. Chen continued, "I need to ensure I have as much information as possible for my meeting. I assume there have been no new developments since we met yesterday. Am I correct?"

"Yes, you are correct. Terri and I know nothing more than we did yesterday. We had planned to meet with Mr. Gregory Brightson later today to discuss John and the case. Do you have a problem with us doing that without you?" I said.

"No, I have no problem with the meeting; however, it is important you keep notes on what is said. If the case remains in Shanghai, which I truly doubt, I will also want to talk to Mr. Brightson and any legal counsel he may consider retaining.

"Now, let's go over the facts as you both know them one more time prior to my meeting."

Ms. Chen, Terri, and I rehashed every point of our discussion of yesterday. Ms. Chen asked several pertinent questions regarding John's relationship with Gregory Brightson; John's wife, Ann; Mr. Dureno; Mr. Da; and Mr. Zhu. Terri and I dug deep to remember any small fact that might be of assistance to Ms. Chen, and after an hour, she announced she was satisfied with the data.

"I believe I am ready for my meeting, and it is getting close to time to depart, so thank you both. I will call you when I return from the meeting."

Once again, we thanked Ms. Chen for her assistance and followed her assistant to the elevator that would take us to the lobby. This time, I spent approximately one half hour enjoying the spectacular art before Terri pulled me toward the elevator to the street.

"Come on, honey, I think it's time we be on our way! This is really not a public exhibit."

"I know," I said. "It's almost impossible for me to rush past without absorbing this. I hate to leave here," I said.

It was hours past lunchtime and a realization we had not eaten lunch hit us when we entered the taxi.

Terri looked at me, and I looked at her. "How about the Golden Dragon again?" I asked.

"Sounds terrific," she said, "but don't you think we might be becoming creatures of habit if this continues?"

"I can't think of a better habit; can you?"

"No, honey, I can't. Let's eat and worry about that later."

I gave the driver our destination, and Terri and I spent the entire taxi ride going over our meeting with Ms. Chen. We were both excited but apprehensive about what her meeting would bring.

We knew this was the beginning of our journey to prove John's innocence.

We arrived back at the Holiday Inn at 4:30 p.m., our appetites satisfied with an excellent lunch. There was a message on our room telephone from Ann Alworth requesting we call her on her cell phone.

Terri returned the call. Ann asked if we would join her and Gregory Brightson for dinner that evening in the Shangri-La Hotel. Ann suggested we meet them at Jade 36 at 8:00 p.m. We agreed.

Our elevator reached the thirty-sixth floor and Jade 36 at 8:00 p.m. The doors opened to a panoramic view of the city, which sparkled with a star-filled sky. A muted red tone accented with cream gave the setting an elegant look. We could see a variety of fresh flowers were scattered throughout the room in vases of various types, colors, and sizes. An elaborate centerpiece had been placed in the center of each table. The maître d met us as we entered the restaurant and asked if we had a reservation. We said no, but that we were supposed to meet Mrs. Alworth at 8:00 p.m.

"Oh yes, Mrs. Alworth. Your names, please?"

"Mr. and Mrs. Watson," I replied.

"Please, follow me."

He escorted us to a window side table for four. The windows ran along the length of the restaurant overlooking the city.

"Mrs. Alworth, may I present Mr. and Mrs. Watson." He bowed as he enunciated our names.

"Thank you, François. Rick, Terri," she said, rising from the table. "I'm so glad you could come!"

Ann gave Terri a hug and me a handshake before turning attention to her left.

"This is Gregory Brightson!"

Gregory Brightson was exactly as John had described him over the past five years. His presence and smile along with his blond hair and penetrating blue eyes lit up the space surrounding him.

Terri whispered quietly to me, "Boy, is he striking!"

"It is such a pleasure to finally meet you both," he said. He extended his hand to me, executing a sincere firm handshake and giving Terri a gentle hug, saying, "May I?" as he did.

Terri replied, "You may," and returned his hug. "It's also a pleasure to meet you."

"Please, join us," he said. He pointed to the two seats opposite him and Ann.

"Unfortunately, we are meeting under difficult circumstances," he said. "Ann has just been describing the situation to me with more detail than I had prior to my trip here. WMT and I will do everything we can to bring this situation to a successful conclusion."

"Gregory, I know you will! The company is so lucky to have someone like you in control!" Ann said. She gave Brightson a very affectionate look.

I was immediately bothered by the corporate speak from Brightson and the gushing response from Ann! I expected different responses from both. John was supposed to be a friend of Brightson's as well as an employee, and Ann appeared to be more interested in impressing Brightson than defending her husband. Maybe I was misreading the conversation. Terri was always better analyzing these situations than me. I would get her reaction on the way back to the Holiday Inn.

"I told Gregory about your meeting with the lawyers in Shanghai and that Ms. Chen was to meet with the police this afternoon. Have you heard from her, Rick?"

"No, we haven't. I really don't expect she will call until tomorrow," I replied. "Mr. Brightson."

"Please, call me Gregory, Mr. Watson."

"Okay, only if you call me Rick, Gregory!"

"Rick it is!"

"Gregory, have you formulated any plan on how you and your company will handle this situation?" I asked.

"Yes, somewhat," he replied. "I discussed this with my in-house counsel as well as Dennis Dureno, and they both agree that hiring a local firm is the place to begin. You know, this is a first for me or my company. We have never had an employee charged with any crime, let alone murder!"

"I believe Ms. Chen will tell us tomorrow the case will be transferred to Changzhou," I said. "I think she believes the Shanghai police will not want anything to do with this one. After all, the murder happened in Changzhou, and Shanghai municipal were only making the arrest at the request of the Changzhou police.

"John asked us here to help prove his innocence," I continued. "It makes sense that we work together and use the same legal firm. Ms. Chen will recommend counsel in Changzhou, if we request it."

"I certainly don't want to work in opposition! Your assistance would be welcomed. We do have some legal contacts here, of course." After a moment of thought, he said, "WMT will retain any counsel she recommends, if the case is moved to Changzhou."

"Mr. Brightson," Terri began.

"Gregory, please, Terri."

"Of course," Terri replied. "I assume you have been in China before? Have you also been to Changzhou and the supplier?"

"I have been in China several times over the course of the project. I made an initial site visit with John and was with him when WMT gave China Universal approval to begin the project. I have developed a fine relationship with Mr. Da, their president."

"I have not personally seen the product but understand from John it is a paradigm shift in water management technology," I said. "The leading-edge industry magazines and other communication outlets have just raved about the product's potential."

Brightson began, "I am not a boastful person, but the design is far advanced in the process of managing how water is stored, controlled, and directed to source of use in a home. The process can also be expanded to offices, hospitals, and other large public indoor places as well, once it is proven in homes, and we fully intend to do so!

"The product reduces cost of operation, saves electricity and water, a valuable natural resource.

"Rick, it is critical John's situation be corrected quickly. I have an obligation to my company and my investors to understand what occurred here and get this product back on track as quickly as possible!

"It's not that I don't believe John, but it's hard to imagine what could have possibly happened to our product that would be that serious.

"We have done many design reviews, preproduction tests, and life tests; so far, none of the tests have shown anything that would lead us to believe there is a problem with the product," he said.

"God help us if John is involved, but if he is, I certainly have a difficult time in believing this is not a personal situation in the plant, not our product!"

Ann replied, "There is no way John could do such a thing, Gregory!" Her face was a cross between annoyance and disbelief. "You have known him for five years. Have you seen anything during that time that would lead you to believe he was capable of murder?"

"Of course not, Ann!" Brightson said. "I was just speculating, poorly I might add, on what could have happened."

During our conversation, the waiter had taken our orders, and the first course arrived at our table.

"Well, I for one am hungry!" I said. "Let's enjoy dinner and finish our conversation over after-dinner drinks, okay?"

"Believe me, Gregory, Rick is always hungry! Right, Ann?" Terri said.

"He has been as long as John and I have known him!"

Brightson laughed. "Okay then, bon appétit!"

Casual conversation ensued throughout dinner and dessert. The French-prepared meal and service certainly met Jade 36's reputation. We adjourned to a bar area tucked into a corner of the restaurant that also provided expansive views of the city below. Shanghai's buildings contradicted the problem of power shortages in certain parts of the city by presenting an extravagant light show every evening.

We had chosen a soft leather cream-colored semicircular couch for after-dinner drinks.

"What are your near term plans, Rick?" Brightson asked.

A small, very attractive young Chinese woman approached, bowed slightly, and requested our drink orders.

Terri began by requesting brandy, and I ordered Irish whiskey on the rocks. Ann ordered amaretto, and Brightson ordered a rusty nail.

"Ms. Chen should call tomorrow morning," I said. "My actions will depend on what happened in her meeting with the Shanghai police. I am hoping she will have the jurisdiction settled, we will know what evidence the police have regarding the murder and John's involvement, and we will be allowed to speak directly with him.

"I'm not sure where to begin until we have that information, "I continued.

"I totally agree. Please, call me when you have talked with Ms. Chen so we can discuss her findings," Brightson requested.

"And please, also, call me, Rick. I have tried to see John several times, as you know, but each time the police would not let me. Now that you and Gregory are here, I am sure everything will be okay!" Ann said. She looked sad.

We said our goodbyes in the hotel lobby with hugs and handshakes and promises to call tomorrow.

The doorman hailed us a taxi as soon as we indicated we were in need of one. He gave the driver our destination, and we were off into Shanghai nighttime traffic.

"Well, that was an interesting evening!" Terri said. We were now into the normal traffic flow of motorbikes, bicycles, and cars.

"Yes, it was. There were times I wondered if Brightson was more concerned about the product launch than helping John, but near the end of the evening, he was more supportive."

I forgot to mention Ann's reaction early in the evening to Terri. We continued to rehash the evening during our ride, coming to no new conclusions, and finally, each of us retreated into the bustling scenery playing out before our eyes as we moved in and out of traffic.

I said my usual prayer regarding Shanghai taxi drivers when we arrived safely back at the Holiday Inn!

Chapter 8

I awoke Thursday, May 15, at 7:00 a.m. to a driving rain pounding on our bedroom window. Terri moaned, rolled over, and asked, "Is it raining?"

I could hear a soft snore before I could answer. I knew after fourteen years of marriage she would sleep for at least two more hours, so I quietly slipped into my running shorts and a Charleston Marathon t-shirt and headed for the gym.

Four miles into an eight-mile run, an aggravated voice quickly brought me back to reality. "Why didn't you wake me up so I could run with you?"

"Sorry, honey, but I know how much you love sleeping in during a rain, so I thought I would get my run in early. I'm hoping Ms. Chen calls this morning with good news."

"Okay, you're off the hook this time, but don't let it happen again!" she said. Thankfully, I recognized the twinkle in her eye. "Besides, if I don't go to breakfast with you, no telling how many trips you will make to the buffet!"

"I know. If I don't control myself, no amount of running will suffice!" I said.

We finished our runs and the rest of our normal morning workout and decided to stop by the executive lounge on the way down to our room to pick up fruit and pastries with juice and coffee.

We arrived back in our rooms around ten in the morning, just as the room telephone was ringing.

"Hello," Terri answered. "Yes, this is Mrs. Watson."

"Oh, I see, and then Ms. Chen will call us early tomorrow morning to discuss her visit with the municipal police?

"I understand her court case will take all day; however, my husband and I would be available in the evening to discuss her meeting if that is acceptable. I see.

"Okay then, we will look forward to her call tomorrow morning."

Terri replaced the telephone on its stand and turned to me. "That was Ms. Chen's assistant. Looks like an unexpected day at court and a special dinner meeting this evening will keep Ms. Chen from talking with us until tomorrow morning."

"Damn it! I really wanted to get everything moving today, especially a visit with John."

"Me, too. I know this will be a big disappointment to Ann, as well. I guess I should call her with the news."

"You call her, and I will call Brightson."

I went into the bedroom to retrieve my cell phone. When I returned to the living area, Terri had already located Ann.

"I'm sorry, too, Ann, but there is nothing we can do but wait until tomorrow morning. Rick is calling Gregory right now to tell him.

"Oh, he is there with you! Rick," Terri said. She held her cell phone beside her. "Gregory is with Ann. No need to call him."

I nodded.

Terri continued her conversation with Ann. "Honey"—Terri's southern upbringing showed through—"we will call you tomorrow as soon as we talk to Ms. Chen. You take care.

"Goodbye." Terri disconnected the conversation.

I said I would call Bin to inform him. We decided the balance of the rainy day would be perfect for reading and relaxing.

Terri and I decided to do room service for lunch about 1:30 in the afternoon. I found after several lunch meetings in the hotel, room service food was actually very acceptable. Besides, it was still pouring buckets outside, and we were into our books.

About two hours later, a good lunch led to head nodding while I was reading when the room telephone jarred me awake.

I reached the phone before Terri. "Hello, yes, this is Rick Watson.

"Hello, Mr. Dureno, yes, I know who you are. Both John and Ann have mentioned you several times over the past five years.

"I understand you are here in China? Both Terri and I look forward to meeting you sometime very soon," I said.

"You're in the lobby? Why don't you come up to our room so we can talk? We're in room 3012. Fine, see you in a few minutes."

Terri was in the bedroom. "Honey, that was Dennis Dureno. He is in the lobby and wants to talk to us. He should be here in a few minutes."

"Wonder what brings him out on a day like this and why he wants to talk to us now?"

The suite's doorbell rang. "I'm not sure, but I believe we are about to find out!" I said.

I crossed the living room and answered the door. Dennis Dureno was the epitome of a marketing vice president. He was at least six feet four with a broad smile and open face that exuded trust. We found out later that he was forty-six and a close friend of Gregory Brightson.

The two had met when WMT was a fledgling startup and Dureno was sold with the product's potential and the opportunity to become wealthy. Brightson convinced him to leave his high-level marketing position with a Fortune 500 company and help chase the dream.

"Please come in, Mr. Dureno, and hand me your umbrella," I said. I placed the wet umbrella in the ornate umbrella stand next to the front door.

"Thank you, Mr. Watson, and thank you for seeing me without notice; I apologize!" he said.

"It is no problem at all. In fact, I was going to call you later this week, anyway."

Terri entered the room. "Mr. Dureno, this is my wife, Terri," I said.

"Ma'am, it's a pleasure to meet you," Dureno said. "Ann has told me so much about you both over the past five years, I feel as if I know you!"

"You are too kind, Mr. Dureno. What brings you here on a day like this?"

"I understand from Gregory and Ann you've come to Shanghai to help John."

"Yes, that is our intention, if we ever get the opportunity to begin!" I said. "I do believe that opportunity will arrive tomorrow, however."

"That's why I'm here. I just returned from the first half of my scheduled visits to potential customers in cities throughout China. I am extremely excited about our opportunity here once the product comes to market. I fully expect I will find the same enthusiasm on the last half of my trip."

Dureno continued, "I am extremely concerned events over these past few days are so serious that they could not only kill the product but the company. I understand John trusts you completely and has confidence you will find whoever is responsible for the murder. Gregory and I certainly need your help to expeditiously prove John's innocence and dispel any claims the product is not ready for the market!

"Mr. and Mrs. Watson, Gregory and I have everything we own invested in this product and cannot afford a failure!" Dureno said.

Panic was deeply rooted in his face. He was no doubt pleading with us to find a solution as fast as possible.

"I fully understand your personal position, along with that of Mr. Brightson's, but our reason for being here is to prove John is innocent, not save the product and maybe the company," I said. I really didn't like his approach or his desire to place the burden of saving the company on our shoulders!

"What makes you believe the murder of Mr. Zhu has anything to do with the product?" I asked. At this point, I did not want to divulge what John had told me during his first telephone call and about the comments Zhu had made to him and their scheduled meeting prior to Zhu's death.

"All I know is Zhu is dead, John is in jail, supposedly charged with his murder, and Gregory is here to decide what to do next," Dureno said.

He continued, "I just don't see how our product could have anything to do with a murder! According to Gregory, we have done everything that is needed to prove our product performs as it is supposed to, and it has passed all quality tests.

"I believe the sooner all of us can prove the murder and John's involvement are separate to the performance of the product, the sooner Gregory will release the product, and we can continue with our plans to take the company public.

"We must find a way to remove any questions about the safety of our product!" Dureno said.

Dureno continued before I could respond to his last statement.

"I just wanted to plead my case to you for finding a solution as quickly as possible. Five years of work and the lives of all of us at WMT are riding on what happens here!"

I was beginning to get upset! I did my best to hold my anger and respond with a civil tone.

"I appreciate your interest and your comments regarding the importance to WMT and its employees in proving John innocent, Mr. Dureno, but let me be clear: we are here first to prove John is innocent. Hopefully, in doing so, we can determine how and if the product is involved."

"That is my fervent hope, as well," he said. "Again, I apologize for arriving unannounced, but I felt I must see you as soon as possible! I will be on my way now. Again, thank you for seeing and listening to me."

I handed Dureno his umbrella and promised to keep him informed of our progress.

Just as he was about to leave for the elevator, I asked, "Will you be available later this week to discuss this situation further?"

"Yes, I plan to be travelling early this week and then working out of the Shangri-La until I return home."

"We look forward to our next conversation, Mr. Dureno," I said. He made his way to the elevator and was gone.

I turned to Terri. "What the hell is going on with that guy? He doesn't give a shit about what happens to John as long as the product ships, and his big check hits!" I was pissed to say the least, and my voice certainly exhibited it!

"Whoa, boy! Don't have a heart attack! We have too much to do!" Terri said. "I don't like the way he came in here and threw John under the bus either! He certainly won't be an ally during this case. He's in this only for himself."

"We need to learn more about this Mr. Dureno," I said. "I'm going to call Gary Hildegard tonight and ask him to do some follow-up for us."

Gary and I never lost contact over the years since we worked together in 1999. He retired from the Marines Corps and stayed in Beaufort with his family. He remained in law enforcement and for several years had been the Beaufort police chief. Gary had access to all the information we would need regarding Dureno. I was sure he would help us.

"We can develop our plan of action once we have talked to Ms. Chen and John. We certainly want to talk to Dureno again!" I said.

"I'm surprised we haven't heard from Ms. Chen," Terri said. "It's almost four."

"Me, too. Do you think I should call her?"

"No, I think we should wait. I am sure she will call as soon as she is able."

I was still offended with Dureno's visit and was pacing the floor, snapping my fingers, a habit developed in the Marine Corps, when Terri shouted, "For God's sake, please stop that pacing! I've reread this paragraph three times! If you're that wound up, go to the gym and work it off!"

"The rain's stopped. I need to get out of here! I need to think! Want to come with me?" I asked.

"No, I'll stay here with my book. I know you think better by yourself in situations like this. Take your cell phone in case Ms. Chen calls."

I grabbed an old, trusty, weather-beaten rain hat and called out "okay" over my shoulder and was out the door heading for the elevator when I remembered I had not given Terri a goodbye kiss. A cardinal sin! I quickly retraced my steps back to the suite.

"It's a good thing you came back, buster, or you would have been in *big trouble!*" she said. I kissed her and once again headed for the door. She was laughing when I closed the suite door and headed for the elevator.

In five minutes, I was sitting at my favorite spot in the train station plaza, watching the people hustle by. Everyone was concentrating on making their departing trains, meeting someone, or finding transportation away from the station.

The giant big screen was playing a commercial on the upcoming visit to Shanghai of the new Simpsons stage play. I was always amazed at the contradiction within China as the country continued to struggle with East vs. West. Capitalism vs. communism!

Young people of working age in China would no longer react to adversity like the older people raised under communism. The influence of Western currencies and the growth of industry to supply the Western world had put more money in their pockets than their parents could ever have dreamed. Only the continued shift west and industry growth would be accepted. A visit to any upscale mall in downtown Shanghai solidified this reality. Couples not just browsing and wishing, but carrying bags containing their evening's purchases.

Older Chinese and the government struggled with this truth, but also recognized the change was too momentous to reverse. Their world had been altered drastically, and society would no longer be completely controllable.

My cell phone rang, and caller ID identified Terri. I hoped she had good news!

"Hello, honey," I answered.

"Sweetheart, Ms. Chen just called and asked if we could meet her in her office at six this evening. She apologized for calling so late, but was in court all day. I told her we would be there at six."

"That's terrific! I'm on my way back to the room, "I replied. "I will be there in a few minutes."

Our taxi dropped us off in front of Lu, Wxi, and Hoa's offices just before 6:00, and the elevator reached the twentieth floor right at 6:00 in the evening. The receptionist stationed in the main lobby said Ms. Chen was expecting us and to go on up to the twenty-first floor. She said Ms. Chen's assistant would meet us at the elevator. We thanked her and started that direction. I lingered a moment to observe a certain piece of sculpture that had caught my eye during our other visits.

"Don't you dare!" Terri said. "Ms. Chen is waiting!"

"I know, I know," I said and followed her into the elevator.

Ms. Chen's assistant was waiting for us at the elevator entrance and escorted us to Ms. Chen's corner office. Once again, she was impeccably dressed in a blue business suit that appeared to be made of silk.

"Good evening, Mr. and Mrs. Watson, I am sorry for the delay in meeting you. I know you are anxious for my report on my meeting with Shanghai municipal and Chief Superintendent Feng.

"Please, let's sit here by the window where it is more comfortable," she said. Ms. Chen lead us to a large conversation area with black leather couches and large black wing-backed chairs that were positioned to see the city view below and her desk simultaneously.

"We completely understand, Ms. Chen, and are grateful for your assistance and guidance," Terri said.

Ms. Chen began, "I met with Chief Superintendent Feng, his senior superintendent of police, Mr. Cheng Gudong, and chief inspector, Que Bei Hua. Chief Inspector Que is in charge of the case and did most of the report.

"I had previously submitted the proper documents required to be recognized as Mr. Alworth's legal counsel, and they were accepted. I was not allowed to meet with Mr. Alworth after the meeting; however, you, Mr. Watson, and I have permission to see him on Thursday morning."

"Finally, that's terrific, Ms. Chen! It's about time!" I said.

"Unfortunately, our system of justice moves slowly and differently than yours. Murder cases are particularly sensitive, especially where an American is involved."

She continued, "It seems there are three critical facts that lead to the arrest of Mr. Alworth. One, his gun was found near the body, which was found in the quality control lab of the plant. Two, witnesses reported Mr. Alworth and Mr. Zhu had a violent disagreement in the quality control lab the day of the murder, and John threatened to kill Mr. Zhou. And third, the quality control lab appeared to have been searched, as though someone was looking for something. Most important, Mr. Alworth's fingerprints were also found on the gun.

"Mr. Zhu's body was discovered by a cleaning lady employee after everyone had left that portion of the plant. To date, no eye witnesses to the murder have come forward."

"Why wasn't John charged based on the preliminary evidence?" I asked.

Ms. Chen replied, "That is the second part of my report. The Shanghai police are planning to transfer this case to the Changzhou police as soon as possible. In fact, the transfer would already have occurred if you and Mrs. Watson had not met with Chief Superintendent Feng last week.

"You mentioned during your meeting that you expected John's company to engage legal counsel in Shanghai within five days, so Feng was waiting for that event," Ms. Chen said.

"At that point, I was not certain what would happen, but I needed to stall for time until Mr. Brightson of WMT arrived in Shanghai," I said. "Looks like it worked in our favor this time!"

"I also requested bail, "she said. "It was denied this afternoon."

"I was really hoping we could arrange bail so John would be able to work alongside us, "I said. "I guess we can also request bail when the case is finally transferred to Changzhou."

"Yes, that would be my advice. The Changzhou court system will have more interest in this case and may act in your favor. There are precedents in other cases involving Americans," she said.

"Is that all the facts the Shanghai police have regarding the murder?" I exclaimed.

"Yes! I do believe what facts they do have are enough to hold Mr. Alworth for the murder!" Ms. Chen replied. She sounded somewhat annoyed at my question.

"Sorry, I did not mean it that way," I said. "I guess I just expected more."

"I believe finding John's gun with his fingerprints on it near the body would be enough for me if I was on this case!" Terri said.

I sheepishly nodded agreement with Terri.

"Where do we go from here?" I asked Ms. Chen.

"You and I meet with Mr. Alworth tomorrow morning at 9:00 a.m. and hear his side of the story," she replied. "In the meantime, I will attempt to find out when the transfer to Changzhou will transpire."

Both Terri and I sat there for a few minutes, deep in thought. "Okay, then we will not take up anymore of your time. I will advise Mrs. Alworth and Mr. Brightson of the results of your meeting and our appointment tomorrow.

"I assume you and I are the only people allowed to see John as of now?" I asked.

"Yes, I believe so."

"I know his wife, Ann, will be greatly disappointed."

"I feel confident after our meeting tomorrow morning, our request for Mrs. Alworth to see her husband will be allowed; however, Shanghai municipal may defer and let the Changzhou police make that decision," Ms. Chen replied.

"We will convey that to her during our conversation."

Terri and I thanked Ms. Chen, said our goodbyes at the elevator, agreeing to meet at Shanghai municipal at 9:00 a.m. tomorrow morning.

Terri and I made our way to street level and hailed a cab to take us back to the hotel.

We were settled in the backseat of the taxi. Terri said, "Honey, please be patient! I know you are used to events moving quickly, but it appears things just aren't that way here." She took my hand in hers while she was speaking.

"Honey, I know that from doing business for so many years in China. It's just that this is personal and all that happened before was business. This time, our best friend's life may be on the line!"

"I understand," she said. "Let's wait until we get back to the Holiday Inn before we call Ann and Gregory. It will be quieter there, and I want to make sure we are saying the same thing to both of them, "Terri suggested.

"Fine with me. I need to call Bin as well and update him on where we are," I said.

Our ride was uneventful, and we arrived back at the Holiday Inn at 8:30 p.m.

"I'm hungry," I said. "Let's have dinner at that little place one street behind the hotel and then make our calls. Remember the name?" I asked.

"You mean Ming's Diner. Remember, the owner moved back to Shanghai from Chicago and opened the restaurant using the same name he used there?"

"Yes, that's it. The food was exceptional the last time we were there."

We arrived in our room about 10:00 p.m. in the evening and decided the lateness of the hour would not be too late for calls. We discussed what we would say to both Ann and Gregory Brightson. I was not as concerned what I would say to Bin, so I called him first. I related the events of the past two days, including our meeting with Dennis Dureno to Bin. I asked him to keep our conversation confidential for the time being. Of course, he agreed. I had always trusted Bin implicitly.

Terri placed a call to Ann, reaching her on her cell phone. I placed a call to Gregory Brightson, but his voice mail was activated, so I left a message for him to return my call.

"Ann, it's Terri. We met with Ms. Chen today, and she gave us the report on her meeting with the Shanghai municipal police.

"Here's what we know. First, John will not be charged in Shanghai, because the case is going to be moved to Changzhou next week. Ms. Chen also said John's gun was found next to the body of Mr. Zhu and that John and Mr. Zhu had a violent argument earlier in the day of the murder.

"Ms. Chen requested bail, but it was denied because of the impending move to Changzhou."

Terri listened intently for a few moments.

"I know you're upset, Ann, but there is nothing we can do until John is transferred to Changzhou," Terri said. Ann was obviously upset.

"I'm afraid there is more bad news," Terri continued. "You will probably not be able to see him until the transfer. Ms. Chen received permission to see him tomorrow, and Rick can be with her. At least we will finally be able to hear John's side of the story."

I could tell from the rest of the conversation Ann did not take the news of not being able to see John well. I could hear her shouting and crying into the telephone. For several minutes, Terri did her best to console her, and it was obvious her attempt was marginal at best.

"Look, Ann, I realize John is your husband, and if this case happened in the United States, things would be different, but we are not home, and things are more difficult in China," Terri said.

It was obvious Ann was lashing out because of not being able to see John. Terri was doing her best to keep her focused.

"Ms. Chen is a very successful and highly respected lawyer here in Shanghai. We are lucky to have her on our side!"

Ann told Terri that she would talk to Gregory Brightson. He would certainly make something happen so she could see John, intimating Ms. Chen had failed.

Terri replied, "Okay, you talk to Gregory!" Terri's voice was controlled frustration. "Maybe he can do something about this situation with WMT's legal connections.

"Rick has placed a call to him, but got his voice mail. Please have him call us when you talk to him. It's important we continue to communicate and work together on this!

"Ann, I'm so sorry things are like they are. We just have to keep working on this. I know Ms. Chen and Rick will have a better understanding of what happened after they meet with John tomorrow morning."

Terri continued, "Let's talk again tomorrow after the meeting with John, and again, if you hear from Gregory, have him call Rick."

Ann agreed.

"We love you, honey! Keep your chin up! Bye for now."

Chapter 9

It was late, and tomorrow was going to be an early and full day. We were both tired and ready for a good night's sleep. I guess because we were both keyed up about the prospect of seeing John the following morning, neither one of us slept soundly. An alarm was not necessary. On Friday, May 16, we were both awake at 6:00 a.m. We dressed, skipped our morning workout, and were finishing breakfast in the Western restaurant at 7:00 a.m.

Ms. Chen's assistant called with a change of plans. Ms. Chen requested we meet her in her office at 7:30 a.m. Saturday morning, May 17, for discussions regarding the process involved to meet with John at Shanghai municipal and what information she believed was critical to get from John.

Now that we knew the case would be transferred to Changzhou, it would be important both of us asked the right questions and took detailed notes to pass along to the new legal counsel.

We had no plans for Friday evening, so after we each had showered and took a quick unplanned nap, Terri suggested dinner at Ming's Diner. I replied, "Are you sure you wouldn't like something different tonight? We have been there at least twice this trip."

"No, Ming's is just fine. I love the food, it's close to the hotel, and Ming does seem to like remembering Chicago when we're there."

"Okay with me," I replied. "That way, I don't need to change into something more formal. My jeans and t-shirt will be just fine!"

"Oh, come on, honey, you look fine in any old thing!" Terri said. She was laughing as if she was full of monkey business as she headed for the bedroom to change.

"Laugh all you want, but you always make me look good when we go out no matter how I'm dressed!" I said.

She shouted from the bedroom, "Flattery will get you nowhere! We're still using your credit card to pay the bill!"

Dinner was superb, and the conversation with Ming made the evening a pleasant one. We returned to our suite about 10:30 that evening, dressed for bed, and quickly fell asleep.

Chapter 10

We arrived in Ms. Chen's office on time Saturday at 7:30 a.m. and quickly began to develop a list of questions to ask John. Ms. Chen described the process to see him.

"When we arrive at the station, we will both be asked for identification and thoroughly searched for weapons and other contraband. The interview room is very small with no wall decoration. It should have one large transparent window the police use to view what happens in the room. All conversations are recorded, and a guard will be with us. He will stand behind John throughout the meeting and no doubt make mental notes of the conversation. You are not allowed to touch John in any way. Not so much as a handshake!

"We will have one hour with him since this is our first visit. If there is a second visit prior to the transfer to Changzhou, it will be thirty minutes.

"I will ask for, but most likely not receive, permission for Ann to meet with John. We will clarify that issue prior to leaving headquarters."

Ms. Chen continued, "We have our list of questions, but don't be afraid to ask additional ones if they come to mind. Please talk to me if you have a concern about a question you are going ask before you ask it. Do either of you have any questions for me?"

We said, "No, we have no questions."

"Unfortunately, Terri, you will be relegated to one of the unpleasant waiting rooms in the station. I trust you have a good book!" Ms. Chen laughed.

"Don't worry about me; I'm prepared," Terri said.

"Okay, let's return to the lobby and secure a taxi for our trip to the police station," Ms. Chen said.

The taxi ride was less than thirty minutes. We arrived at Shanghai municipal a few minutes early for our 9:00 a.m. appointment.

Ms. Chen spoke briefly to the sergeant at the reception desk. The sergeant instructed a constable to escort us to the security area for inspection prior to our going into the working area of the building. The sergeant told Terri in English where the main waiting room was located. We said our goodbyes. The sergeant watched her as she made her way to the designated waiting room.

The security check was as Ms. Chen described it. We emptied our pockets and briefcases of everything except writing objects and paper. The constable informed us where we could retrieve our belongings after our meeting. Ms. Chen said later recording equipment was permitted with prior written request and written return consent, but only with the accused's legal counsel present.

The top three floors of the building were devoted to interview rooms, short-term holding cells, and cells for longer-term inmates. All the interview rooms were on the third floor. I understood preventing jail break attempts was the thinking in putting cells on the top two floors.

The constable escorted us to the elevator located in middle of the third floor along the west wall. It was the same one Terri and I used in our initial visit. We followed him to an interview room in the back of the building. The constable handed us off to the constable responsible for overseeing movement in and out of the interview rooms.

Ms. Chen thanked both constables for their assistance and completed the required forms. Ms. Chen indicated where I was to sign, and she followed suit.

The interview room fit Ms. Chen's earlier description. It was small, square, and painted battleship gray. One half of one of the walls was filled with a two-way glass used for outside viewing, as Ms. Chen said. The chairs were aluminum with black imitation leather seats. The table was made of aluminum, approximately six feet in length, with a one-foot partition running the length of the table. There was one very bright light fixture made up of three florescent lights hanging over the center. A single lamp was sitting on the table, but was not lit. There were no decorations on the wall. There were, however, instructional leaflets of various types.

Ms. Chen had reiterated her instructions on behavior in the interview room during our taxi ride to Shanghai municipal.

She added there would be no exchange of any documents or other items, nor any movement from our chairs. We would have our allotted one hour from the time John entered the room. Ms. Chen reemphasized most constables who worked on the top three floors spoke adequate English.

We had been seated on one side of the table in the interview room for only a few minutes when the back door opened. I was positioned on the right, and Ms. Chen was on my left.

John entered with the constable immediately behind him. He was dressed in a red jumpsuit with black cloth slippers. The slippers had very thin rubber soles. It had been five days since the night the police had taken John from his room in the Shangri-La. He looked tired, but alert.

John sat, positioning himself in the middle of his side of the table. He spoke first to me. "It's about time you got here! What did you take to get here, a slow boat to China!"

"Look, if you had paid me the hundred bucks you owe me from our Clemson/Carolina football bet, I would still be at the lake fishing, and you would be on your own, buddy!" I replied. "Where's my money!"

Ms. Chen did not know what was happening, and her normal placid face had an almost horrified look. John and I laughed. I can't ever remember any conversation between us starting on a serious note! Ms. Chen sighed with relief, probably thinking we were nuts!

"How are you?" I asked.

"As well as expected, considering the circumstances. How is Ann? Have you talked to her, and who is this with you?"

I said, "First, we have talked to Ann several times since we arrived, and she is eagerly waiting for her first visit to see you. Second, this is Ms. Chen. She is a partner with the law firm Lu, Wxi, and Hoa. Bin Wang recommended them and Ms. Chen as the best firm and counsel in Shanghai to help us."

John tuned his gaze toward Ms. Chen. "Thank you for helping me, Ms. Chen. I know if you were recommended by Bin Wang, you must be the best!"

"You are too kind, Mr. Alworth. I am privileged to assist you in any way I can," Ms. Chen said.

"Gregory is here, also," I said. "I have talked to him about your call to me, and he and I have agreed to work together.

"I also had a fairly long conversation with Dennis Dureno regarding the murder of Mr. Zhu and the effect on WMT," I said.

John replied, "I don't know if I can trust either one of them anymore! I'm afraid getting the new product introduced and into production is more important to them than helping me prove my innocence!"

"You know them best; however, Gregory seems highly concerned about not only the product but proving you innocent. I would tend to agree with your assessment of Dureno. He paid a visit to Terri and me at the Holiday Inn yesterday expressing his concerns regarding the product launch, tying it to his and the company's financial security," I said.

"We can't let the product go to production until we know what Zhu knew and have corrected any problem he found," John said. The tone of his voice showed his concern. "WMT cannot be held responsible for causing deaths to users of our products! Rick, we have got to stop them from going ahead until we have the data!"

"I agree, John," I said.

I continued, "John. We have one hour with you today and a lot of questions to ask, plus we want to hear your side of the story, so let's get started. Can you tell us what happened beginning with your arrival in Shanghai last week?"

John began, "What day is it?"

"It's Saturday, May 17," I replied.

"Ann and I arrived here on May fifth, twelve days ago. On May sixth, I went to the plant, and Ann went with me. She had to meet a client in Changzhou.

"We took a late morning train. The train arrived on schedule at 11:30 a.m. A driver from the plant was waiting for me. Ann had hired a car and driver for the day. We agreed she would meet me at the plant if her meeting ended prior to my train departure time.

"The ride from the train station takes one half hour. I was to meet Mr. Zhu for lunch. We decided to have lunch in the company cafeteria to save time. After eating, we went to the QC lab to review the data from the last series of tests on the new product.

"We required UCPC to run seven normal life tests, and then we extended the time parameters for additional safety."

John continued, "I was primarily interested in the extended life tests. Those tests are critical, along with the rest of the data, for final sign-off from UL and the other government agencies. We had successfully completed all the required tests, except the extended life tests. They were the last piece of the puzzle. All the skids had been greased, and we expected final sign-off within days after our tests were submitted.

"We had also been working with the state governments and knew they would fall in line once we had approvals at the federal level. Once we had approval, WMT could begin advertising as well as proceeding with the public offering," John said.

Ms. Chen asked, "What happened during the review of the data?"

"Mr. Zhu and I had many conversations about the life tests. He wanted to run another series, increasing the extended life test time significantly. He planned to use a model UCPC had developed to accelerate the test speed but still produce the acceptable data. I disagreed. In fact, during our last conversation, I ordered him not to perform those tests! I never could get him to tell me why he felt the need to run additional tests.

"Our engineers had run the same life tests in house in conjunction with the tests at UCPC. Our engineers had already gone past the extended life test time and found no changes occurred from the initial extended life test. We were completely satisfied we had met the required standards. We found nothing abnormal from our additional testing."

Ms. Chen asked, "Did Mr. Zhu run those additional tests?"

"I'm afraid so!" John replied. "Mr. Zhu told me he had run an extended life test using new sample number eight. He told me he had done so as soon as we arrived in the QC lab. I asked him why after I explicitly told him not to run any additional tests.

"His reply was he knew more about the product than anyone, even WMT's engineers, and would not open up his company to possible catastrophic consequences if the product failed in the field or caused harm to a customer.

"I was not happy and made it very clear he had no right to perform an extended life test on another sample without approval.

"Once I knew the data existed, I was obligated to review it, add it to the data package, and make any corrections the data may have uncovered.

"An additional life test might also give UL and the government reason to delay approval and request additional extended life tests on all seven samples. Obviously, this could be fatal to our schedule for both the product and going public.

"WMT has a huge amount of capital invested in this project. We can't afford a long delay in approval and introduction."

"Did you argue with Mr. Zhu?"

"Sorry to say now, but you could probably hear me all over that part of the plant!"

"Witnesses say you threatened to kill Mr. Zhou; did you?"

"Rick, you know how hot I can get sometimes! I really don't remember what I said. I was really pissed at him! But damnit, his stubbornness could cause WMT to fail!

"But for God's sake, I didn't get mad enough to kill him!" John said. He was showing the strain of being in a small cell for five days.

"What did Zhu say to you regarding the extra testing on the eighth sample?" I asked.

"That's the rub. He refused to give me the data or talk about it. He said it was too sensitive to discuss there and wanted to meet me at the Shangri-La later that evening to discuss it.

"I wanted to meet with Mr. Da, the president of UCPC, to discuss the extra tests, but Zhu begged me not to. I had calmed down by then and decided it would do no good to throw Zhu under the bus until I knew how bad the problem was."

John continued, "If his data was right and the product could cause serious illness or death to customers, I needed to understand the test data and immediately get the information back to WMT."

"What happened the rest of the day?" Ms. Chen asked.

"We reviewed the rest of the data I had requested. I asked several report format changes be made and ready the next day when I returned. I had planned to return to Shanghai on the early evening train, and then return to Changzhou the next morning."

"Why did you do that?" Ms. Chen asked. "Why not just stay overnight in Changzhou?"

"Ann wanted to come back to Shanghai that evening, and I agreed to go back with her, so I was not prepared to stay in Changzhou."

I asked, "Did you talk with anyone else while you were there?"

"Just the normal people. I always do a walkthrough of the plant to make sure the initial production is running smoothly. I talked to several of the supervisors and the plant manager. There was the normal flow in and out of the QC lab after our heated discussion."

"Did you have your briefcase with you the whole time you were in the plant?"

"No, I usually leave it in the QC lab."

"Was your gun in the briefcase?" I asked.

"I assume so. I have carried that old gun in that briefcase for years. Always legal, of course. When we first began to look for suppliers outside of the United States, some of the countries were not as safe as you might think. Having my gun with me made me more comfortable. Funny, though, I never needed it in all my travels.

"It was a pain sometimes to get permission to carry it, but I always made sure I had the proper paperwork," he said.

"What type of a gun is it, Mr. Alworth?" Ms. Chen asked.

"It is a Smith and Wesson Chiefs Special, Model CS45."

"Your gun was found next to Mr. Zhu's body. Any idea how it got there?" Ms. Chen asked.

"Hell, no! I certainly didn't leave it there. Come on, Rick, you know I could never kill anybody. I had enough of that in 'Nam!" John said. He was becoming more stressed during the conversation.

"Did you notice anyone lingering near your briefcase while you were in the QC lab?" I asked.

"No, just the normal flow, like I said before. Qian Xin was in and out several times."

"Who is Qian Xin?" Ms. Chen asked.

"Xin, his English name is Billy, is Mr. Zhu's right hand guy. He does all the testing and writes the original report," John replied.

"Did Billy know about the extended life test using sample eight?" I asked.

"I'm sure he did. There wasn't much that happened there without his involvement!"

"But you did not talk to him about the additional test?" I asked.

"No, the conversation, argument, between Zhu and me happened when we were the only ones in the office."

"Mr. Alworth, do you have any idea who might have wanted Mr. Zhu dead?" Ms. Chen asked.

"Not really. You know, it's hard to truly know the dynamics of any plant, especially in China, if you are not there daily. Mr. Zhu always seemed to be in control and have the respect of everyone when I was there," John replied. "Why? You think someone took my gun from my briefcase and killed Zhu to frame me?"

Ms. Chen replied, "Anything is possible. I believe you when you said you did not kill Mr. Zhu, so we need to find out who did!"

"Did you have any conversation with anyone from WMT regarding the additional extended life test and Mr. Zhu's concerns?" I asked.

"Dennis Dureno happened to be in the plant when I arrived. I didn't know he was going to be here, but he was in the area making calls and decided to stop by to see what was happening with the product.

"I told Dennis Zhu had run an unauthorized test, and we had a big problem meeting the schedule. He asked if I had a copy of the test data. I told him no, that Zhu refused to give it to me but wanted to meet me in Shanghai later that night to discuss the data.

"I told him this could be the ultimate show stopper!"

"Do you have any idea where Mr. Zhu kept the data in question?" Ms. Chen asked.

"I suppose in his desk in the QC lab. He kept everything of importance there."

Ms. Chen asked, "Did you infer how serious the outcome of this test was and how long the schedule could possibly be shifted?"

"Yes, I did. Again, I told Dennis the whole project could be in jeopardy!"

"And what was his reaction?" Ms. Chen continued.

"Mostly panic! He ranted and raved about how the company couldn't afford for this to happen: how we needed to get the data and refute it so the project could continue as scheduled."

"How did you react to his attitude? Did you agree with him?" Ms. Chen asked.

John replied, "I told him to calm down! This was not the place to have this kind of discussion. I told him I would meet with Mr. Zhu that night, get the data, and we would proceed based on what I found out."

John continued, "I told Dennis this product had to be right, that I would review the data and call Gregory. Dennis wasn't happy, but reluctantly agreed."

"Did you talk to anyone else about this before leaving?" I asked.

"Only Ann. She came back to the plant as I was about to leave for the train station."

"Did you tell her about the additional tests and their importance?" Ms. Chen asked.

"Of course."

"Where did you meet her when she came to the plant?" Ms. Chen asked.

"She met me in the QC lab."

"Isn't it unusual for her to be in the plant?" Ms. Chen asked.

"No, not really. She has been all over this plant with me several times," John said.

"What about Mr. Dureno? What happened to him? Did he leave before or after you left the plant?" Ms. Chen questioned.

"He left before. He had another call to make and was going back to Shanghai on the late train."

"So," I said, "you and Ann left the plant and returned to Shanghai?"

"I went back," John said. "Ann stayed in Changzhou. She said something had come up, and she needed one more day there."

"Is that normal? I mean Ann changing her schedule?" Ms. Chen asked.

"I wouldn't say it was normal, but it happens enough that I didn't think anything about it."

"Did she have personal effects with her for an overnight stay?" Ms. Chen asked.

"She normally carries a small emergency bag with her when she travels," John replied.

"And she had her bag with her that day?"

"Why do you ask?" John said.

"No particular reason," Ms. Chen replied. "Just seems a bit unusual to me."

"Yes, she did," John answered. "Wait a minute, you don't think Ann would have anything to do with this, do you!" John asked. "That's just not possible!"

I quickly came to Ms. Chen's defense. "I don't think Ms. Chen believes Ann had anything to do with the murder; she was just asking a normal follow-up question."

I asked John, "What time did you leave the plant, and which train did you take back to Shanghai?"

"I left the plant sometime after four in the afternoon, after spending a full day with Zhu and really getting nowhere, planning to catch D301 at 6:33 in the evening to arrive in Shanghai at 7:47 that night."

"You said planning to catch D301 at six thirty p.m.," Ms. Chen said. "Did you make the train?"

"No, I was so mad I needed some time to think, so I went to Wei Yuan Park, just a short walk from the train station."

"Did you talk to anyone while you were there, did the driver see you leave the station area after he dropped you off, or did you purchase anything near the park?" Ms. Chen asked.

"No, I just sat by the small lake on a bench, thinking about what to do next, trying to determine what was so important that Zhu felt he needed to meet me in Shanghai and how to handle him when he arrived," John said.

"What train did you take back to Shanghai?" Ms. Chen asked.

"I ended up taking the seven twenty-one and arrived on time in Shanghai at eight forty-three."

"I'm sure you can prove you were on that train?" Ms. Chen asked.

"I still have the stub, if that's what you mean?"

"Please don't lose it. Mr. Alworth, you told us you had a fairly violent argument with Mr. Zhu regarding the additional extended test, right?" Ms. Chen asked.

"That's right."

"Based on that, did you expect any trouble when Mr. Zhu arrived at your hotel, and why do you think Mr. Zhu wanted to meet at your hotel?" Ms. Chen asked.

"No, I really did not expect trouble," John said. "Yes, I was hot, and we argued, but we finally agreed to disagree and to stop arguing after Zhu requested he come to Shanghai to see me."

"Why Shanghai and your hotel?" I asked. "Why not somewhere in Changzhou?"

"I offered to stay in Changzhou and take the last train back to Shanghai, but Zhu said no, he wanted to meet in my hotel."

"Any idea what was so important or secretive that Zhu felt he could not discuss anything near the plant?" I asked.

"No. I have been searching my brain since this happened, but can't find an answer," John said.

"What time were you supposed to meet with Mr. Zhu?"

"He told me he would take D5404 at eight seventeen p.m. and arrive at nine forty-six p.m. and be at the hotel by ten," John said.

Ms. Chen asked, "Did you talk to Mrs. Alworth prior to leaving on the train?"

"No," John answered. "I did not."

"Is that normal not to talk to your wife in a situation like that?" Ms. Chen asked.

"No, not really. We normally talk late at night during trips."

"Anything else you can think of about the day or the events?" Ms. Chen asked.

"No, that's it," John said.

I looked at my watch and noticed our hour was quickly coming to an end. I looked at Ms. Chen. "Have we missed anything we discussed in our meeting this morning?"

"No, we have covered all the topics," she replied.

"I really want to get into the product, but our hour is almost up. It looks as though we will have to wait until our next visit.

"Ms. Chen, do you know when the transfer to Changzhou will happen?" I asked.

"They're going to transfer me to Changzhou!" John said.

"Yes, Mr. Alworth. Shanghai municipal will not handle this case. They arrested you on a request from Changzhou. Final jurisdiction will be there."

"That's just fucking wonderful!" John exclaimed. "Sorry about my language, Ms Chen! So that means a new lawyer and beginning again!"

"Yes, Mr. Alworth, but I know several very capable firms in Changzhou that handle these types of cases, and I will meet whoever in engaged to ensure they have everything I have when the case is moved."

"Is Gregory aware of the transfer to Changzhou?" John asked.

I replied, "Gregory is fully aware of what has happened and the transfer to Changzhou.

"He has agreed to cooperate completely with Terri and me. He has agreed to pay all the costs occurred during the investigation. He seems very concerned about you and proving your innocence, as well as protecting the product, but he has assured me your safety comes first!"

"I sure hope so!"

The constable pointed to his watch and motioned to Ms. Chen that it was time for us to leave and for John to return to his cell. Ms. Chen asked for just one more question. The constable agreed.

"Mr. Alworth, who is the best person for Mr. Watson and I to talk to regarding the product before our next meeting with you?"

"Talk to Dennis Dureno. He has all the data you will need."

The constable made a move to take John back to his cell, and we rose and prepared to leave the room.

John asked on his way out the door, "When can I see Ann?"

Ms. Chen replied, "After your transfer to Changzhou!"

We could not hear John's comments as the door closed, but knew they were laced with profanity!

We waited a few seconds to make sure John was not coming back into the room and left.

Terri met us as we walked toward the door. "How did it go?" she asked.

"Not bad," I said. "We asked all the questions we had on our list and more. Ms. Chen, where do we go from here?" I asked.

She replied, "First, I need a few minutes with Chief Inspector Que, and then back to my office. We can review our notes and observations."

We exited the building into brilliant sunlight and the synchronization of busy Saturday Shanghai morning traffic. Terri and Ms. Chen were chatting while I attempted to hail taxi when Terri excused herself to Ms. Chen and said to me, "I almost forgot, Gregory called you on my cell while you were with John and wants you to call him."

"Okay. Did he say what he wanted?" I asked.

"He wanted us to have dinner tonight with him, Ann, and Dennis Dureno. He didn't say, but I assume he wants to discuss what happened today with John and our plans going forward."

"Did he mention anything to you about another lawyer or John's possible transfer to Changzhou?" I asked.

"No, he just inquired about dinner."

One of the famous almost red Shanghai taxis responded to my wave and pulled to the curb. Ms. Chen gave the driver directions to Lu, Wxi, and Hoa. A normal thirty-minute ride was fifteen minutes longer because of the late morning rush hour. We arrived at the building at 11:00 a.m.

We spent the next hour in Ms. Chen's office going over our visit with John and compiling the data from our notes. Terri's cell phone rang during our meeting. Terri apologized. "I really must take this one," she said. "It's Ann."

She returned several minutes later. "Ms. Chen, I'm sorry about the call; I thought I had put my phone on vibrate."

"That's no problem. I am sure she is anxious to hear what happened with John."

"Yes, she is!" Terri said.

"I told Ann we were with Ms. Chen, going over your notes from your visit with John, and asked if I could call her back later. She said she just had one question for now and that was when could she talk to John?

"I relayed Ms. Chen's information about John's Monday transfer to Changzhou. Then Ann asked would she be able to see him on Monday.

"I said I didn't know when, but promised to bring her up to date tonight at dinner."

Ms. Chen said, "I believe Ann should be allowed to see him in Changzhou no later than Wednesday of next week."

"Thank you, Ms. Chen. I will tell her that tonight."

Ms. Chen spent a few minutes reviewing our work. She said, "I believe we have included everything that happened during our talk with Mr. Alworth. This should give your new counsel an excellent base to begin Mr. Alworth's defense. I talked to Chief Inspector Que when I excused myself just before we left Shanghai municipal. He informed me Mr. Alworth would be transferred to Changzhou Monday morning."

She continued, "I think it is important you or Mr. Brightson secure counsel quickly based on the transfer timing. Lu, Wxi, and Hoa will be most happy to recommend someone if you wish. As I said before, there are several excellent firms in Changzhou."

"Thank you for your assistance, advice, and offer to help us in securing counsel," I said. "Your guidance so far has been invaluable. I'll discuss our next move with Mr. Brightson tonight at dinner and call you tomorrow," I replied.

"May Rick and I buy you lunch?" Terri asked. "It's just after twelve thirty. You do take lunch, don't you?"

Ms. Chen replied, "Well, normally it's something small that, believe it or not, I bring from home. An old habit from my days just starting out, however, today can be an exception!"

Ms. Chen recommended a small local restaurant just around the corner from her building. It was a very warm afternoon, but the walk was leisurely, and we enjoyed mingling in the constant flow of other businesspeople going to and from their lunches.

Ms. Chen explained as we walked that Wang Bao He was Shanghai's oldest eating establishment, founded in 1774. It was noted for its Shanghai hairy crab, but Ms. Chen said it was very expensive and really not worth the price. In fact, she said, it was not worth any price to her!

We laughed at her comment as we entered the restaurant. Every table was taken, and there was a sizeable group of businesspeople waiting for the next open table. Ms. Chen made her way to the podium in the front of the seating area, said a few words to the young lady doing her best to organize the crowd, most with cell phones stuck in their ears, and made her way back to us.

"It will only be a few minutes," she said.

True to her word, the young lady motioned to Ms. Chen within five minutes and led us to a table in the center of the room. Several diners stopped Ms. Chen to say hello on our way to our table.

"For someone who doesn't do lunch often, you certainly command respect here," I said.

"I really don't do lunch that often, but our firm has a small ownership position here and most of the individuals are government officials or attached to the legal profession or the courts," she said. Ms. Chen was obviously being very low key about her importance.

"I recommend any crab dish on the menu, just not the hairy crab!" she said, laughing.

Terri and I ordered crab soup as a starter and deviled crab, another specialty of the restaurant. Ms. Chen ordered a bowl of crab soup. She also ordered a bottle of Shaoxing wine as an accompaniment to our lunch. Shaoxing was noted as a special wine only served by Wang Bao He.

Everything we ordered was as advertized, and we thoroughly enjoyed the wine!

We walked back to Lu, Wxi, and Hoa with Ms. Chen. We thanked her for allowing us to buy her lunch and her assistance these past few days. I agreed to call her Monday morning with the details and decision from our dinner meeting that evening.

It was 2:30 in the afternoon when we found a taxi to take us to Nanjing Road so Terri could do some shopping.

Nanjing road is a pedestrian-only broad, straight avenue at least two miles long and contains one of the most popular shopping areas for Shanghai locals as well as national and international tourists. Everything from McDonald's (several) to Gucci can be found somewhere along the street.

Several McDonald's are very distinct in Chinese design, along with the customary golden arches, and occupy two stories. The menu is mostly conventional American sandwiches with additional products specific for China. A Big Mac is a Big Mac is a Big Mac!

We meandered in and out of shops of all kinds, just enjoying comparing Shanghai prices to US prices on similar items and examining the quality of goods produced locally. Terri found what she was looking for, and we spent the rest of our time watching the extensive diversity of people shopping and browsing. Terri and I always enjoyed watching the mostly petite, wide-eyed Chinese children as they marveled at all of the trinkets being sold by street vendors or the large, colorful motorized train that carried passengers in multiple cars from one end of Nanjing Road to the other.

I called Gregory Brightson during our taxi ride to Nanjing Road. We agreed to meet at the Big Bamboo Sports Bar and Grill at 7:00 that evening. Gregory said he had enough Chinese food for a while and wanted a cheeseburger from Bamboo!

It was 5:00 that afternoon when we arrived back at the Holiday Inn. Just enough time to rest and freshen up for what Terri called our Shanghai gourmet dinner!

Big Bamboo Sports Bar and Grill was located behind the Portman Ritz Carleton Hotel. It was not a trendy place, but one noted as a casual friendly place in Shanghai for an American to hang out, catch up on sporting events in the United States, drink, and eat burgers. I had enjoyed their American-style weekend brunch, as well as their wonderful cheeseburgers, several times while staying in Shanghai.

Gregory, Ann, and Dureno had arrived prior to us and had located a large booth in the left corner of the restaurant. They waved us over to the booth when we walked into Big Bamboo's front door.

US seventies and eighties music slapped us in the face as we entered, and it was clear the mostly Western crowd appeared to be happily into the music, beer, and food.

Gregory and Dureno rose to greet us. "Hey, guys, welcome to Big Bamboo!" Gregory said as we approached the table. It was obvious this was his kind of place.

Ann remained sitting, but said hello and made room for us in the booth.

"I hope you like this place," Ann said. "Both Gregory and Dennis are frequent visitors when they're in Shanghai."

"I really have to come here at least once per trip," Dennis said. "I do like Chinese food, but I need a cheeseburger fix every time before I leave."

A young male waiter saw us arriving and was at our table just a few seconds after we sat down. "What can I get for y'all?" It was obvious he was an American, probably from the South.

I scanned the table and saw Ann, Gregory, and Dennis were drinking Buds.

"I'll have a Corona with lime, please," Terri said. I followed suit. The waiter smiled, spun, and was on his way to the bar to enter our beer order.

"Have you ordered?" I asked.

"No, we waited for you, but I am definitely ready!" Gregory said. He was looking at Ann and laughing.

Our beers arrived. Gregory ordered a second round, and we all placed our dinner orders. It was cheeseburgers all around. No one shied away from any topping. Each order was all the way!

Dinner arrived, and we were almost finished enjoying our terrific American-style-all-the-way cheeseburgers, fries, slaw, and beer when Gregory began, "Rick, how did it go today, and where do you believe we should go from here?"

"Ms. Chen, Terri and I developed a set of questions to ask John during the interview. Ms. Chen and I were able to ask all of them and more.

"Ms. Chen also met with Chief Inspector Que on Wednesday to ask him what evidence the Shanghai police have tying John to the murder of Mr. Zhu," I said. "According to Chief Inspector Que, there were two incriminating facts regarding John's possible involvement in the murder. One, John's gun was found near the body with John's fingerprints on it, and two, John and Mr. Zhu had a violent shouting match in the afternoon of Mr. Zhu's death. Witnesses say John threatened to kill Mr. Zhu during the argument."

Ann became very emotional during my comments. She almost shouted, "That's just bullshit! We all know my husband would never kill anyone! He isn't capable of doing such a thing!"

"Look, Ann, you and I know that, but people here in China don't know John as we do. They have a murder case on their hands, and I believe they will do what is easiest to find him guilty!" I said.

"Okay, okay. Let's get back to what we do to prove John is innocent and get the product into production," Gregory said. "Is that all they have on him?"

"I made that exact comment last week to Ms. Chen, and she politely said, isn't that enough!" I replied.

Terri joined in. "First, we need to determine what Mr. Zhu was going to tell John that night. It may give us a clue to why someone murdered him. Also, once John has been transferred to Changzhou, we need to find a good defense counsel and establish a working relationship with the Changzhou police."

"I completely agree," I said. Gregory nodded his agreement as well. Ann was still seething from my comments, but nodded in agreement. She said, "All I know is I want to see him as soon as I can!"

"Do you have the list of firms in Changzhou that Ms. Chen provided?" Gregory asked.

I replied, "Yes, I do." I handed the list containing six firms to Gregory. Ms. Chen had indicated her preference in priority order.

Gregory perused the list before asking, "Can we trust her judgment in rating the firms on this list?"

"I certainly believe so from our work with her over the past few days," I replied.

"You agree, honey?" I looked at Terri for her comments.

She said, "I would agree. Ms. Chen has been fantastic!"

"Then I will contact the Changzhou, Huaide, and Zhanglinfang law firm and arrange a meeting with them on Monday to discuss John's case," Gregory said.

I looked at my watch. It was just past 8:30. "Let me call Ms. Chen on her cell. "I know it's late and a Saturday evening, but she said I could call her anytime if I needed something. Let's see if she meant it!"

I placed my call and Ms. Chen answered on the second ring.

"Ni hao."

"Ms. Chen, ni hao, this is Rick Watson. Please forgive me for calling so late on a Saturday night, but you said I could call anytime if I needed something. Well, I'm taking you up on your offer!"

I wasn't sure Ms. Chen fully understood what "taking you up on your offer" meant, but she didn't hesitate with her reply. "Mr. Watson, I know you would not call me at this hour unless it was very important."

I hoped she wasn't angry. It was difficult at times to determine if someone from China was angry. The culture was to always be polite, but I forged ahead anyway. For me, this call was very important.

"I'm here with Terri, Ann Alworth, Dennis Dureno, and Gregory Brightson. We are discussing John's transfer to Changzhou and your prioritized list of law firms to help us. Since John is being transferred on Monday morning, we would like to be in a position to see him in the afternoon. To do that, we need to engage a firm to work with us.

"Mr. Brightson has selected, and we all agree, Changzhou, Huaide, and Zhanglinfang from your list as our first choice to work with. Do you think it would be possible for you to call them and arrange for Gregory to talk to them this evening?"

"I believe that is possible, Mr. Watson. I will call immediately and return your call after I have spoken to Mr. Xue, president of the firm."

I thanked Ms. Chen and told her we would await her call.

"We might as well hang here until she calls back," Gregory said. "Anyone for another round?"

"Sounds good to me," Dureno replied. It struck me that this was the first words he had uttered during the evening. Strange!

Our next round of beers arrived. I was about halfway through my bottle of Corona when my cell rang.

"Ni hao. Oh, ni hao, Ms. Chen, were you able to reach Mr. Xue?"

Ms. Chen explained that she had talked to Mr. Xue, and he was waiting for us to call him on his cell number. He was at an art exhibit and reception for his daughter in downtown Shanghai, but would be most happy to talk to us.

"Xie xie, Ms. Chen. We all really appreciate you doing this for us! I will call you Monday morning to discuss transfer of our data to the new firm. Xie xie again, you have been most helpful."

Ms. Chen gave me Mr. Xue's cell number, and Gregory placed his call immediately upon my hang-up.

Evidently, Mr. Xue's cell phone did not ring on Gregory's end, so he was not sure the call went through. "Ni hao, ni hao. Is this Mr. Xue with Changzhou, Huaide, and Zhanglinfang?"

It was.

"Ni hao, Mr. Xue, my name is Gregory Brightson, chairman of Water Management Technologies. I believe Ms. Chen, from Lu, Wxi, and Hoa has called you within the last half hour on my behalf. Am I correct?"

Mr. Xue obviously answered in the affirmative.

"I apologize for a call so late on a Saturday evening, but I have an employee who has been charged with the murder of one of our suppliers' employees. The firm, Universal China Production Company, is located in Changzhou. My employee was arrested in Shanghai at his hotel by the Shanghai Municipal Police Department, but he is being transferred to Changzhou for trial.

"I need a firm to work with us in solving this case and proving his innocence plus defending him in court if necessary," Gregory said. "Ms. Chen said you would be her first choice to work with us if you had time."

Mr. Xue asked a few questions regarding the case and replied his firm would agree to represent us. Gregory scheduled a mid-morning meeting on Monday.

"Xie xie for being gracious and taking my call. Again, I apologize for a call at this hour. I will see you at ten thirty Monday morning. Please enjoy the rest exhibit and reception."

"*Zai jian*, Mr. Hue." Gregory hung up and faced the group with a smile on his face. "Now we can really get this thing in high gear!"

We all laughed and toasted his comment!

Ann asked me, "Do you know what time John will be transferred to Changzhou on Monday?"

"Ms. Chen thought it would be before lunch time. She believed it may be possible see him later that day," I replied.

"Then I guess I had better not miss my meeting on Monday morning with Mr. Xue and get his firm signed up as our lawyer!" Gregory said. "Rick and Terri, I assume you will accompany me?"

"Yes, we will. I will arrange our train tickets and call each of you over the weekend regarding the schedule.

"Ann, Dennis, will you be going with us?" I asked.

Dennis said, "I would like to, but I have a very important meeting with a potential customer here in Shanghai Monday morning."

Brightson said, "That's right, Dennis. You keep working on new customers!"

Ann said she would be there whatever time we gave her.

I asked Dennis if he could spend time with Terri and me on Sunday afternoon to describe the new product to us. I explained we did not have enough time to ask John to describe it on the day we interviewed him.

He agreed and suggested we meet him at 2:00 in the afternoon at the Shangri-La. The lesson, as he called it, would take about two hours, allowing for discussion time.

We finished our beers, commented about the excellence of the food, and departed Big Bamboo agreeing to do this again!

It was a beautiful evening. Not to hot with a warm breeze waffling across this portion of the city. We asked our taxi driver to drop us at People's Square. This would be a perfect night to stroll the length of Central Boulevard running through the square and to enjoy sitting and watching people.

I believed Shanghai is second only to Hong Kong in the light show proudly presented nightly, and People's Square was the perfect location to view the office buildings on the Pudong side of the Huangpu River. China's leading architects had designed a series of very unusual buildings along and just off the river. The buildings were special during the daytime, but nighttime lights of various colors made them spectacular for viewing.

We selected a bench on the corner of Central Boulevard and Huangpu Road across from the Shanghai Grand Theater. According to the huge lighted billboard in front of the theater, Shakespeare's *A Midsummer Night's Dream* was playing this week.

The theater's expansive white marble steps lead to massive center front doors made of polished brass. The center doors were flanked on either side by smaller doors also made of polished brass. The lobby was magnificent with its centerpiece chandelier and surrounding smaller chandeliers sparkling like diamonds just out of reach for plucking!

Watching people after several beers relaxed us sufficiently enough to head back to the Holiday Inn. This past week had been stressful, and Terri and I decided we were ready for a day off! Our normal schedule contained a daily workout and a long run. We had missed several days the past week, and our bodies felt like it.

We walked to the opposite end of Central Boulevard and hailed a taxi back to the Holiday Inn.

Our home away from home was a welcome sight when we arrived just after midnight. It did not take long for us to complete our bedtime preparations, do a final recap of the day's events, and fall asleep with the television playing softly in the background. Terri had developed a habit of watching television to induce sleep. Me, I was always gone a few seconds after my head hit the pillow!

Chapter 11

The Sunday, May 18, morning sun streamed through our bedroom window waking us at 8:15 in the morning. I could not remember the last time either of us had slept this late. I looked out the window and gazed down on the morning flow of vehicles and pedestrians going about their Sunday morning business.

It was going to be another beautiful day in Shanghai, and we were going to do our best to enjoy it. Terri had ordered our picnic basket when we arrived back in the hotel Saturday night. All we had to do this morning was shower, dress, pack a small bag with a change of clothes, have breakfast, and call for the transportation she had also arranged that would deposit us at Dongping National Forest Park.

It was 9:30 when the black Buick limousine arrived to take us to the park for our half day of relaxing!

We had visited Dongping before and knew to deposit our picnic basket and change of clothes in a locker located in the bath house of the pool.

The first business of the day was stretching and running fifteen miles. Since we had not maintained our normal running schedule the past week, today's run would be leisurely so we could fully enjoy the scenery around us. The flowers were bursting with a cornucopia of colors, shapes, and sizes. We stopped along the way to enjoy the multitude of fragrances that captured our noses.

We ran around the largest lake in the park, impressed by the number of boats on the lake and enthralled by the panorama of exercisers, dancers, sidewalk artists, and families with small children out for a day of enjoyment.

Our run complete, we jumped into the pool to cool off. We retreated to the bath house to retrieve our change of clothes and our picnic basket when our after-run stretching was completed. Terri always made fun of me when I said no one needed a shower after a run if a chlorinated pool or lake was available!

The balance of the morning was spent lying on our blanket sipping flavored water, finishing the last of our picnic lunch and watching people. A few brave souls approached us to ask if we were Americans. They were anxious to practice their English while they learned where we were from, why we were here (no, we did not tell them we were here to solve a murder!), and our thoughts on China.

All in all, we considered the first half of the day a success as we piled into the backseat of our black Buick limousine for the ride back to the Holiday Inn.

On the ride back, I was struggling with the thought that John Alworth may have committed a murder. I had known John for thirteen years and considered him my best friend. You know, we meet many people in life we call friends; however, only a few actually become true faithful friends. John was that to me. He was the brother I never had!

There had to be something more than what we knew to date. I know the circumstantial evidence and John's gun at the scene was damning; however, I was struggling with a motive for John to commit such a heinous crime.

There was just no way this could have happened the way the police believed it did! Not John Alworth, but who else would have a motive?

What could be so critical to the launch of the product that made someone want to kill Zhu? Clearly, the WMT engineers and John believed the product was ready for launch. What did Zhu discover, and how do I find it? Where was the information now? I hoped it was still in the filing system in the QC lab or in Zhu's desk, as John thought.

I was extremely anxious to begin working with the police in Changzhou and to get my hands on anything I could find in the plant.

Ms. Chen had already talked with the commissioner of police in Changzhou about my relationship with John and his desire for me to participate closely with the investigation team. He had agreed to full cooperation.

At 1:00 in the afternoon, the doorman hailed another one of those famous nearly red Shanghai taxis, gave the driver the address of the Shangri-La hotel, and we were off. The hotel was a good hour ride from the Holiday Inn. It wasn't long before each was once again enthralled with everything around us, so the drive felt quicker than we assumed it would. We arrived at the Shangri-La at 2:00 in the afternoon for our "lesson," as Dennis called it. The meeting was arranged for a small meeting room on the second floor of the hotel.

The escalator in the main lobby took us to the second floor, but the meeting room was in the west section of the building. Our path to this section of the hotel took us past the extremely long succession of restaurants featuring various cuisines from around the world.

We found the meeting room and Dureno at the same time. He was walking down the hall from the opposite direction, simultaneous to our arriving in front of the meeting room door.

"Hey, guys, sorry I am a bit late. I had planned to have the room more organized and ready for our lesson, but my wife was up very early and called me with a problem at home."

"You're right; it is early in Charlotte." *What time is it there?* I asked myself. I looked at my watch. It was four in the morning there. "I hope everything is okay?"

He replied, "Everything is okay. My wife sometimes gets panicky when I have been gone for a long period. Today, night was one of those times."

"Okay," he continued. "Let's start learning about the product." The meeting room featured a large mahogany conference table with ten comfortable leather chairs surrounding it. There was a full photographic product display hung across one wall of the room, a projector and an independent laptop sitting at the end of one side of the table. The table also held draft brochures of the product aimed at future customers. Dennis took a few minutes to load his laptop and project his first slide on the screen at the end of the room.

Dureno said, "My plan is to first show you an introductory DVD we have been using with customers to announce the coming new product, and then go into detail on the product features, hard materials used producing the product, software development and features, along with product process flow detailing how the product works in the home and the product's benefits.

"Is this what you would like to see?" Dureno asked. "Does it hit the mark?"

Terri and I looked at each other. I could tell by the look on her face that the presentation would be more than sufficient. I replied to Dureno,

"I believe you have all the data we need to understand the product and its uses, as well as how it functions.

"We need a product understanding to determine the starting point when we finally get to work in the plant and have the ability to follow up on any ideas that may arise toward proving John innocent," I said.

Before Dureno could begin his presentation, Terri asked, "Before we start, I thought Ann might want to join us, but she didn't answer her cell phone. Have you seen or talked to her?"

"No, I haven't, but I had breakfast with Gregory this morning, and he told me she was visiting a client today somewhere on the other side of the city. Gregory also sends his apologies. He said he would try to join us later this afternoon. He had several things from back home to finish."

"Thanks, sorry to interrupt," Terri said.

"No problem," Dureno replied.

"Okay, then, let's get started," Dureno said. "First, I need each of you to sign the confidentially agreement in front of you." We placed our signatures in the proper place on the form and handed them to Dennis, who thanked us for signing.

Dennis continued, "You both need to understand the concept of the product and the process to use it. This slide presentation will accomplish that."

Dureno turned on his computer and brought up his presentation. "This is the same presentation I have made to selective potential distribution customers over the past several months. For obvious reasons, no presentation has been made to the media.

"We call our product 'Master Clear Water Filtration System.' You will see why as I proceed.

"We have developed a very simple process to install your own water management system in your home. The system consists of small point-of-use water tanks at locations throughout the home that traditionally dispense water. For instance, your bathroom showers, your bathroom sinks, your kitchen sink, toilet tanks, and various special sinks that may be scattered throughout a home. Plus any specialty water-dispensing locations, like a sink located at a home bar.

"Our system does away with any handles to turn on water, both hot and cold. A homeowner has several options to activate the flow of water. All options begin with a touchpad that replaces a faucet handle. A pad similar to your iPod or touch telephone pad.

"If the homeowner desires hot water, then he or she simply pushes the temperature desired. For cold, push that desired temperature. The homeowner also has the option to push an event on the touchpad, such as shaving, washing your face, washing your hair, and so forth. If this option is chosen, the system will energize a predetermined setting for the process chosen. Similar to pushing the pad on a preprogrammed touchpad selection on a kitchen microwave. The homeowner also has the option do his or her own programming as well.

"The tank below the sink, for instance, is normally two and a half feet high and three feet in diameter. We have designed state-of-the-art proprietary heaters and chillers that are inserted into the tank to either rapidly heat or cool the stored water in each individual tank. Our software does the controlling work."

There were drawings, pictures, and schematics scattered throughout the presentation that were very helpful in understanding the product, both hardware and software.

Dureno looked at Terri and me to make sure we were following his explanation to this point. He continued, "What I have described so far is advanced and as a standalone product would revolutionize the industry; however, what I am going to describe next is so far outside the accepted norm you will be stunned by its simplicity and dynamics.

"The second half of the product is a water recirculation and processing system that mirrors what happens in every water treatment plant in every city and town in the world!

"City or well water is fed into a holding tank approximately six feet tall and eight feet in diameter stored in a basement, garage, or maybe a laundry room. The main tank dispenses water to the small point-of-use holding tanks as water is dispensed. This action is controlled by our software. Waste water is also fed to another section of the tank to be chemically reprocessed into fresh water ready to be reused.

"Solid waste that is filtered from the used water drops into a holding area where it is dried and burned at an extremely high temperature. The residue falls into a sanitary bag that the homeowner removes weekly and places in the normal trash pickup.

"Simple, but brilliant! No wasted water. Water that is so pure no community water-processing system can match it. The homeowner only has one job: take out the waste!

"Here are the costing models that show annual savings on water bills and system purchase payback times in various parts of the country as well as how much longer a conventional well will last because of a significant reduction in pumping from the main water source using this process.

"The water retention rate is above 93 percent. That means a homeowner will use very small amounts of fresh water throughout the year. The system will allow the homeowner to adjust that flow rate if so desired.

"You can imagine the impact on water usage throughout the world and the reduced need for large water treatment plants when this product is made available to large venues!" Dureno said. "Third world countries will be able to provide water quickly and efficiently to outlying areas that now struggle to obtain pure water. All that is needed is a source to obtain water.

"Okay, that's a start. That was a lot to throw at both of you, so I'll give you a few minutes to catch your breath, and then we will go through any questions you may have," Dureno concluded.

"I need something to drink." Dureno asked, "May I get each of you something?"

Terri wanted a bottle of water, and I asked for a Diet Pepsi. Dennis left the room, and Terri and I stared at each other for a few seconds. "Do you realize what we just saw!" Terri said.

I was stunned at the potential of the product and began to realize how someone might take desperate steps to ensure this product would ship on time. I had no idea what other companies were doing in this area, but felt WMT was far ahead of anyone else. John was right about this product going to market and that WMT going public would make several people millionaires several times over!

"I had no idea this was what John was talking about when he told us about what he was doing in China. My God, this is huge!

"That presentation just whet my appetite. There is a whole lot more we need to know about the product," I said.

"I agree," Terri replied. "Why don't we develop a list of questions for Dennis so we don't forget any?"

We began to develop our list for Dureno while he was out of the room getting drinks. We had completed it and were reviewing the brochures and other data sitting on the table when he returned. "Here you go," he said as he handed us our drinks.

"I'm sure you have questions about the product and process, so let's get started."

I began, "Again, Dennis, this is fantastic. I had no idea WMT's product was so far advanced from today's products. I haven't read or seen anything about the development of this product. How have you been able to keep this a secret until now?" I asked.

"We have confidentially agreements like the one you all signed with each person working at WMT, and we have the same for all our suppliers. We also require each potential distribution customer sign before we do the presentation."

I asked, "What about your potential distribution customers in China? As you know, Western businesses have had significant problems with confidential data ending up in the wrong hands here or having your product duplicated."

"No doubt!" Dureno said. "We are extremely careful who we discuss this product with here and only present top-level data during our presentation. We have found that some development areas and councils are taking legal responsibility more seriously today, so we try to identify the best ahead of time and meet with the local government officials to request their support.

"Fortunately, we are a small company with a small market share that produces normal water management products just like you see in Home Depot or Lowe's, so the press and the industry analysts don't pay us much attention. Of course, being privately held is a tremendous advantage, and luck has been on our side!"

Terri asked, "Have you applied for any patents?"

"Yes, we have, and they are close to being approved. We are able to operate under the patent-pending clause until we receive full approval," Dureno replied.

"I know from talking with John this last week, all the life tests have been run, and the product has passed with flying colors," I said. "Will I have access to all that data so I can do my investigation?"

"Of course," Dureno said. "I will personally talk to Bill Wilkens, our VP of engineering, and Walt Douglas, our head of design engineering to make sure they give you everything you need."

"Thanks, I will probably need that data sometime late next week when I hope I can begin in the plant.

"I know generally how the process works from your presentation, but how do you ensure the water continually meets standards?"

"Good question. The water treatment system is controlled by our priority software. There is an intricate monitoring system that monitors and reports real time individual water standards data to a control panel located on the front of the tank. This panel will sound an alarm or, in a worst case scenario, shut the system down if the process goes out of specification."

"So, what happens if there is a shut down?" Terri asked. "How will the homeowner maintain water supply?"

"If for some reason the system must be shut down, the software automatically goes straight to city or well water from the original outside source. There are several fail safes to assure catastrophic failure will not shut off outside fresh water."

Dureno continued, "Each system has universal monitoring software that notifies WMT customer service and engineering when there is a failure in a customer's system. Customer service will call the customer and arrange a service call. Engineering, along with quality control, will maintain trend data on failures and recommend specification and production changes to manufacturing as they arise.

"The customer will receive a monthly e-mailed report detailing the health of the system and process just like the one you receive from OnStar if you drive a General Motors car or truck."

Terri continued, "You mentioned the product will be sold to homeowners through distributors; am I correct?"

"That's correct. Distributors and big box retailers."

"How will a distributor who sells your product to a homeowner be kept in the loop with events that happen with the product?"

"We want the distributor or big box retailer to become our partner in selling this product, but WMT will remain the main contact with the end customer. It is important WMT has control of quality and the system, not the middle man.

"Of course, we will continue to communicate, train, and make marketing suggestions to our initial customers to assist them in improving results."

"You said the waste is burned and falls into a sanitary bag for weekly disposal," I said. "Two questions. One, what happens if the waste generated needs dumping sooner than one week and two? How does the customer know the waste is being dealt with effectively, that the water remains clean?"

Dureno said, "An alarm will sound if waste is being generated at a higher level than normal to alert the customer to change the waste bag. The main control panel displays this data on a real time basis for the homeowner.

"We will also offer a remote duplicate monitoring control for placement anywhere the customer wishes as an optional feature. The customer will also have the option to allow us to monitor and make the waste bag change for a nominal fee and a yearly agreement.

"Our final safety features are controls that will shut down and lockout the system if the customer chooses not to follow the procedures for waste disposal.

"This is just one reason we will highly recommend and push for each customer to sign up for the monitoring system and waste removal program."

"Have you established a price for the product and system?" Terri asked.

Just as Dureno was about to answer her question, Gregory Brightson entered the room. "May I answer that question, Dennis?"

"Yes, sir, Gregory," Dureno replied.

"Sorry if I interrupted the presentation, but I was here in the hotel and wanted to see how things were going," he said.

I moved away from the table to shake Gregory's hand. "You are certainly not an interruption. It's always a pleasure to see you."

He hugged Terri and took a seat at the table across from us. "I believe we have developed an outstanding program for our distributers, big box retailers, and finally, the end customer. A distributor or large retailer will have several options. The system can be purchased in varying volumes and discounts based on those volumes, or a distributor can sign up for a specific quantity of systems and maybe receive a larger discounted price or just purchase one at a time. Of course, that is the highest price, and we would not recommend that option.

"The end customer has the usual methods of payment as with any other large-dollar purchase. However, there is also a lease option available to the customer as well. That will also be facilitated through the distributor or retailer. We will work with both parties in that transaction."

He continued, "We have also been working closely with the federal government and several state governments to offer the customer a tax incentive to purchase and install the system. We are very excited about this particular option. We believe, with the proper advertising, this option will be our largest generator of sales!

"We will also require a software license no matter if a customer buys or leases and to pay a yearly software maintenance fee."

"So, back to my original question," Terri said. "What is the price of the system?"

Gregory responded, "Basic model, $10,995 retail price."

Dureno jumped into the conversation. He said, "That's just the suggested retail price, but here are some neat opportunities. With the tax incentive of $5,000 from the federal government and an additional $2,500 from specific states, the cost actually is about $4,500 to the buyer, and if you are a volume builder, we can reduce the cost further through marketing incentives. A large-volume builder can install our system in each of his or her houses for under $2,000! Potential huge sales for the builder and a gold mine for WMT!"

"Now you know why this thing with John could not have come at a worse time!" Gregory said. "Here we are ready to launch, still have the cat in the bag, and have a product that is endorsable across the world as the most innovative green product of our times!

"This crap could be devastating to WMT. I have to determine very quickly if we really do have a potential problem with the quality of our system, or if this is a personal issue between Zhu and John."

"Wait a minute!" I said. "You really don't believe John had anything to do with Mr. Zhu's murder, do you?"

"No, no, of course not," Gregory replied. "I should have said a personal issue between Mr. Zhu and the killer!"

"All I know is our window for launching this product is tight, and we can't wait weeks to make a decision!" Dureno said. His voice showed a hint of desperation.

"I understand, Dennis!" Gregory replied. It was obvious he was annoyed with the comment from Dureno. "I agree we must act quickly, but again, I also will not be responsible for selling a product that may have lethal implications to one of our customers. WMT has a fiduciary responsibility to our board as well as a moral obligation to our customers. I have no problem making the decision. I just need more data! Hopefully, we will begin to sort things out on Monday in Changzhou."

It was time to end the session and the conversation regarding the current situation. It was plain there was a problem between Dureno and Brightson. More discussion would not be productive.

"Dennis, thank you very much for your time today," I said. "I'm not sure what to say about the product other than it is truly incredibly innovative, and I agree it may be the most innovative product of our times.

"Terri and I understand how critical it is to launch this product on time; however, may I remind you we are here to prove John did not murder Mr. Zhu or anybody else! We will certainly cooperate as much as we can to help you meet your dates, but we also need to understand the position of the police as well."

"You're absolutely correct," Gregory said. "I promise by late tomorrow morning I will have an agreement with a counsel in Changzhou and be ready to meet with the Changzhou police."

"Guys, it's after five," Dureno said. "I don't know about you, but I'm hungry. Anybody for an early evening dinner at Big Bamboo's?" He was smiling sheepishly. "I haven't had lunch!"

"That's fine with me," Gregory said. "Let me call Ann's room and see if she would like to join us."

"We're in," said Terri. "Our lunch seems like a long time ago!"

Gregory reached Ann in her room. She agreed to join us in the lobby at 6:00 p.m. per Gregory's suggestion.

Dennis said, "I need some time to clean up this room and take everything back to my room. Why don't I meet you in the lobby at 6:00?"

"I need sometime also," Gregory said. "See you at six. Will you and Terri be okay until then?" he asked.

"We'll be fine," I replied. "I'm sure there is a Corona somewhere in the hotel with our names on it!"

"No doubt," Gregory said. "Rick, Terri, please don't think I don't appreciate your help here. I do! I am confident working together we will get to the right answers and find the real killer and launch the product on time!"

"We appreciate your confidence in us. It won't be easy, but we don't have any choice, do we?"

"No, we certainly don't," Gregory replied.

"Dennis, do you need any help taking things back to your room?" Terri asked.

"No, I got it. See you in the lobby."

We went our separate ways at the bank of elevators just outside the meeting room, Brightson and Dureno heading up and Terri and I heading down. Our elevator was situated on the outside of the building, and we could see as we rode down the afternoon had been a good one.

The lobby bar was busy. A large birthday party had taken place in the hotel during the afternoon, and guests were beginning to arrive for two wedding receptions. We managed to find a seat after doing a quick walkthrough. It was always fascinating to watch families and friends gathering for any Chinese celebration. The polite interchange between couples, older people, and the younger set. Elders still commanded a special sort of respect in the culture, and everyone abided by it.

Being in Shanghai was like being in New York when it came to fashion. The young women of Shanghai were the leaders in Chinese fashion. Currently, stiletto heels with short tight dresses of vibrant colors were in. The young men favored very modern designed suits more colorful than anywhere in the United States or Europe.

The older generation mostly still clung to the tradition of mix and match Western-style clothing. They had no problem mixing plaids and stripes along with paisley. Something you would never see in America. Of course, traditional Chinese dress was always acceptable. Anything Mao, however, had been discarded some years ago by all but the ultra-faithful.

We ordered Coronas with our usual lime when the young lady in a tight red dragon oriental dress with a provocative slit up the side saw my wave from across the room where she had just deposited drinks to a table of wedding reception guests.

"You know, I still don't have a comfortable feeling about Dennis," I said. "There is just something about him and his almost desperate approach that worries me! Did we ever hear back from Gary Hildegard about him?" I asked Terri.

"I had forgotten about that one!" Terri answered. "I'll follow up with an e-mail when we get back to the room."

I nodded in agreement.

The player piano was emitting a string of 1970s tunes from famous US crooners like Frank Sinatra, Tony Bennett, Jack Jones, and Dean Martin when we spotted Ann entering the main door of the hotel at 6:10 p.m. She caught sight of us when she was halfway across the lobby, waved, and hurried to our table.

"Thank goodness you are still here!" she said. "I was afraid you all would be gone, and I wasn't sure where we were going for dinner. My cell connection was bad, and I didn't hear where Gregory said we were going."

Terri said, "We are going to the Big Bamboo."

"I should have known!" Ann said. "Those two profess to be fond Chinese food, but 99 percent of the time, we end up at Bamboo's!"

"I'm sorry I missed your meeting with Dennis, but this little business of mine isn't eight to five, five days a week! The young women who come here with their husbands aren't as adventuresome as we were. Most of them require a considerable amount of hand holding!"

She continued, "How did it go?"

"It was very informative and educational," I replied. "I had no idea the incredible impact the product will have on not only the United States, but potentially the rest of the world! What we learned today is a solid baseline to begin our investigation next week in Changzhou."

"Speaking about Changzhou, do we know any more about my husband's transfer and when I can see him?" Ann asked. "This not being able to see and talk to him is driving me crazy!"

Terri said, "That's something to discuss with Gregory at dinner tonight. Last information we got was that John was going to be transferred Monday."

"Ann, I'm sorry I didn't offer to buy you a drink when you arrived," I said. "Would you like one now?"

"No, I'll wait until we get to the restaurant."

"Probably a good thing because they are both coming across the lobby as we speak!" Terri said.

Gregory gave Ann a hug and smiled at us. "I'm sorry we're late. Dennis and I had a telephone call we had to take."

"Everything okay?" I asked.

"Absolutely!" Gregory replied. "Just day-to-day business issues. Are we ready for dinner?"

We all agreed we were and headed out the main door of the hotel for the short walk to Big Bamboo's. Dennis and Ann took the lead and appeared to be deep in conversation all the way to the restaurant.

Big Bamboo's was busy as usual; however, Gregory motioned for the owner, and after a short conversation, we were seated in the exact table we sat last time we were there.

"Do you always sit at the same table?" Terri asked.

Gregory replied, "I try. I like the position of this table." He pointed to each item as he spoke. "We have a perfect line on a big screen that carries ESPN, you can see everything that is happening outside the restaurant, and you can't miss who enters the front door!"

We all laughed and agreed. I was not sure why he would be interested in who was entering the restaurant, but did not have enough interest to ask.

We each ordered the exact same meal we ordered the last time we were here, cheeseburger plates all the way with beer. No reason to change when what we had last time was perfect!

Gregory began, "So, Ann, how has your day gone?"

"Fine! I was telling Rick and Terri, these young women I am working with who are here for the first time are certainly not adventuresome like we were at their age. They seem afraid of their shadows! They want me to make decisions for them on the simplest things! I guess it's because most of them come from well-to-do families who never gave them the opportunity to fail!"

"Ah, but that puts money in your pocket!" Dureno said.

Ann replied, "Just like a marketing man! Always looking for the next angle!"

"Okay, okay, let's not pick on poor Dennis," Gregory said. He was using his best sympathetic voice while quietly laughing.

"I for one would like to have more discussion on the product," I said. My question was directed at Gregory. "John mentioned Mr. Zhu had performed an unauthorized additional extended life test, and that was where the trouble occurred. Any idea why he would want to do that and what he may have found?"

"I have no idea! We have performed life test after life test throughout the development of the product with no problems. Every time we made a hardware or software change, we did new first-piece inspections to the change and in some cases ran new PPAPS.

"We continue to run baseline engineering and quality analysis as part of our project management tool. Our stage gate process includes fixed stopping points and official approval sign-offs before going on to the next stage. Any discrepancy anywhere in the product or system would be found and dealt with before moving on!

"I believe we need to determine if Mr. Zhu changed the parameters of the test when he ran the additional extended life test on sample number eight and why!" Gregory said. "And if he did, why did he not inform John he was making the change with his reason for making it?"

"That's an excellent point, Gregory, and certainly something we need to discuss with John when we see him next week," I said. "However, I still don't see why the additional test would be a reason to murder him. Surely anything he found could be corrected. Maybe not on the current launch schedule, however."

"I hope you are correct about that! WMT and I personally have 90 percent of our joint capital tied up in this basket. I just find it hard to believe there could be something so egregious in the system after all the work we have done over the past three years!" Gregory exclaimed.

"I think this whole thing is bullshit myself," Dureno said. "There is nothing wrong with the product or the system that manages it. Someone had a grudge against Zhu and used the product as the cover-up!"

"That's some statement, Dennis," I said. "What data leads you to believe that?"

Dennis continued, "Look, if I wanted to cover up my actions, all I would need to do is pick up the gun from John's briefcase, do the deed, and leave his gun there as the perfect piece of evidence. Hell, everybody in the plant knows John always had his gun with him and where he kept it!"

"So you believe findings during the additional extended life test is a ruse to confuse us and an act as a cover-up for murder!" Terri said.

"Absolutely!" Dennis said. "I think we talk to John, do another run with our engineers in Charlotte, ensure there are no surprises in the data, and release the product for production! Then, Rick, you and Terri could concentrate on finding out who really killed Zhu and free John."

"Based on what you just said, you believe Zhu really did not run the additional test himself, but gave the responsibility to someone on his staff, and that someone falsified the data, handed it to Zhu, and waited for sparks to fly," Terri said.

"Exactly! The killer knew if the police bought it, they would chase their tails constantly, missing the real facts. I believe someone wanted Zhu dead, but not because of the additional test. In fact, I believe the killer could give a shit about our product and its launch. His only interest was to confuse the police and us."

"Then how do you explain leaving the test data there for all to see and analyze?" I asked.

Dureno said, "Well, you might have a point there, but that's why you guys are here to seal these types of holes in theories like mine!" He threw his hands up while laughing at this point. Not sure what to say next.

Gregory responded, "I wish it were that simple, Dennis. While I agree with you it is critical we release the product to production and launch, and I have a hard time believing we haven't done our homework, I must positively ensure we have a safe product."

Our cheeseburgers arrived, and we spent the balance of the dinner talking about mostly nothing! After we finished our meal and the bill had been settled, Ann asked Gregory what the plans were for Monday morning.

"I contacted Mr. Xue as Ms. Chen suggested, and he has agreed to a nine thirty appointment tomorrow morning," Gregory said. "I plan to take an early morning train to Changzhou. I had not planned to stay in Changzhou, but take the train back and forth each day I needed to be there. It's only an hour's ride."

"Terri and I have made no arrangements to go to Changzhou. We were waiting for you to complete discussions with Mr. Xue. I will call Mr. Da and hopefully arrange a meeting with him at his plant sometime Monday afternoon. Do you think you will you have things wrapped up with Mr. Xue by that time?"

Gregory said, "I certainly plan to. I want to make sure someone from Mr. Xue's firm is assigned to John's case and has contacted the Changzhou police to notify them they are representing John and to ensure John has been transferred to Changzhou before I return to Shanghai. I will call you on your cell when negations are complete."

"I think Terri and I will also return to Shanghai on Monday evening. If everything goes as planned, we will check out of the Holiday Inn on Tuesday morning and set up our base of operations at the Grand Hotel in Changzhou city center. I've stayed there several times and always found it to be a high-quality hotel."

"Gregory, when do you think I should be in Changzhou so I can see John?" Ann asked.

"Tuesday will probably be the earliest, but I will know better after meeting with Mr. Xue. I will call you as soon as I know anything so you can make your plans. I know how anxious you are to see him."

"You're right about that! Please, push our new lawyer to do what he or she has to do to make Tuesday happen. I don't think I can bear it much longer!" Ann said. She was definitely in a state of high anxiety!

"Do you want me to reserve your seat on the train for tomorrow?" I asked. "I'm going to do our reservation when we get back to the hotel."

"Yes, you may. I looked at the schedule, and there is a train at 7:19 that arrives at 8:37. Mr. Xue will have his driver waiting for me when I get there. I believe the train number is D30."

"I'll take care of it. First class cars have aisle and window seats. Which do you prefer?"

"Makes no difference. Do you think there will be a problem calling late tonight?" Gregory asked.

"No, I don't think so, but I will call you if there is a problem," I said.

Terri asked, "Dennis, what is your schedule for Monday?"

"I have rented a day room at the Grand in Changzhou and plan to spend the rest of the day catching up on e-mails and telephone calls," he said. "I also believe I have a conference call late Tuesday afternoon our time with our sales managers in Europe."

We worked our way to the front of the restaurant. Gregory spent a few minutes with the owner, thanking him for our table and promising to return soon. Terri and I asked the hostess where most of the taxis were this time of night. She suggested Fuxing Road, a major cross street on the way back to the hotel.

Terri and I said our goodbyes when we reached the corner of Hushan and Fuxing Roads, once again instructing Gregory to secure his ticket at the will call window at the train station. It would be a first class ticket on D30 reserved in his name. I also reminded him that the taxi ride from the Shangri-La could take over an hour that time of the morning, and the station would be packed.

Once again, our suite looked extremely inviting upon our return to the hotel. I personally was going to miss all the room. We would not pay the extra cost for a suite at the Grand, but did reserve a large king room somewhere on the executive floor. Having no additional charge to access both the executive lounge and the gym was a lifesaver!

I called the hotline number for train reservations as soon as we arrived in our room. No problem booking the D30 train for Gregory. His assigned seat was25, a window seat in the first class car number 1 with a return at 9:00 that evening. We chose D5424, a noon-time train that would arrive in

Changzhou at 1:25 in the afternoon. Our first class seats were also in car 1, mid-car in seats 22 and 23.

Neither of us was in any way ready for bed, so we spent about one half hour reviewing the evening's conversation, especially the comments made by Dennis Dureno. Terri remembered she was going to e-mail Gary Hildegard and did so. I hoped we would hear something from Gary in the morning. I still had an uneasy feeling about Dennis Dureno. I couldn't put my finger on it, but I knew I would. We turned the lights out about 1:00 in the morning.

Chapter 12

Tuesday, May 19, arrived with bright sunshine flowing through our living room window at 6:30 a.m. We had forgotten to close the blind when we went to bed. Terri rolled over to face me and suggested the light streaming into the suite was my fault since I was the last person to come to bed. I put up a halfhearted defense, padded into the living room, and closed the blind.

"Now you close the blind!" Terri said. "A lot of good that will do when I'm awake!"

Not wanting to create more discussion, I quickly gathered my workout clothes, opened the bedroom window blinds, and scurried for the bathroom! A pillow hit the door just as it was closing!

We stretched, ran eight miles, and enjoyed the sauna followed by breakfast in the executive lounge. That left us two hours to pack, shower, dress, and check out. Our train to Changzhou was scheduled to depart at 12:04 p.m. Large amounts of luggage were impossible to handle on trains in China, so we hired a car to drive ours to the Grand Hotel. If everything went as planned, our luggage would arrive at the hotel about the same time we would.

I sent an e-mail to Gregory from my Blackberry asking if his tickets were at will call earlier that morning. He was on the train when he responded, saying everything was fine, thanked me for reserving his tickets, and that the next stop was Changzhou. So far so good!

It was 11:15 a.m. when we departed the hotel for a five-minute walk to the train station. Most of the early morning throng of office workers, students, craftspeople, and tourists had already successfully managed to navigate their way through the labyrinth of temporary aluminum dividers that wound around in front of the station like a corn field maze, through security and up the stairs into the main waiting room.

The enormous LED screen centered above the entrance to the train station was twice the size of my favorite screen at the opposite end of the plaza. We maneuvered our way to the entry door of the station along with a small group of travelers. As usual, there was a significant amount of pushing and shoving as we approached the entry and security. I am never sure if the pushing and shoving is the result of travelers being late for their trains or just Chinese culture! Fly any in-country Chinese airline, and my observation will be confirmed. Flight attendants make the traditional announcement regarding remaining seated until the plane comes to a complete stop at the gate; however, Chinese passengers of all ages completely ignore these warnings by standing and removing their belongings from overhead bins immediately upon the sound and vibration of the plane's wheels making first impact with the runway or ignoring any unloading protocol by pushing their way up the aisle to be the first person to deplane.

We placed our hand luggage on the belt of the security tunnel that moved past a security camera manned by at least two security agents who were conveniently busy talking and mostly watching the female travelers rather than the screen. There was a pedestal for wanding those passengers who failed the magnetometer, but no one was manning that position. We retrieved our bags and decided the escalator directly in front of us was the right one to take us to the main waiting room.

The room was overflowing with travelers of all ages and sizes carrying packages, small luggage, and cardboard boxes. Containers of food to be consumed on the train were also prevalent. Most Chinese travelers preferred homemade snacks to those sold from carts on the train. The large train schedule board located at the end of the waiting room detailed every train arrival and departure time. An official-sounding voice announced each train. That announcement started a barmy scramble to be the first passenger to descend down the stairs to the train platform.

The waiting room seats were small black leather couches each designed to accommodate three travelers with three aluminum trim rails on each side of the couch and one rail that ran along the bottom. It was obvious they had serviced the traveling public well, but were past the time to be replaced. The tile floor had been ugly from the beginning and now was barely considered clean, and the trash cans were overflowing.

As usual, there were no seats, so Terri and I spent a few minutes walking through one long shop that occupied one wall of the waiting room. The shop contained drinks of all kinds, food stuffs (most of which we would never purchase), souvenirs, and over-the-counter medicines.

Our 12:15 p.m. train was announced while we were inspecting the herbs and medicines in the store, so we made our way to the exit door. Again, pushing and shoving was in play as we slowly made our way to the front of the line to have our tickets verified and punched by the attendant.

Our train loading point was on the opposite side, so we followed the crowd across the overpass and down the stairs to the platform below. It was already a warm day when we exited the stairs onto the platform. The train side platform floor had car number positions stenciled in large white numbers. We made our way to the area designated for car number one. We could hear the horn from the train approaching the station as we arrived.

China is a world leader in train transportation. Their trains varied from the older, slow trains with hard bench seats and unclean bathrooms, to the most modern fast trains with first class airline-type seats, ultra-clean bathrooms, and free bottled water.

Our train was one of the new fast trains that reached speeds of 200 miles per hour. It was white with a sleek rounded aerodynamic nose designed to break any friction created by air movement in front of the train. There was very little sound other than a whoosh of air moving past us while the train was stopping.

We found our seats and settled in for the one-hour ride to Changzhou. The train left on schedule, and the cityscape gave way to country side scenes. The train followed a river for a significant portion of the trip. Several other rivers connected to the main river along our route. These rivers flowed past the small villages and towns inland from the main river.

All the river boats we observed were wide and somewhat flat to accommodate the low bridges that crossed each river. These boats carried small loads of goods and many times were the home of the owner. Very large barges with similar designs carried all types of heavy products.

New roads and bridges were being built all along the river in varying stages of completion. I had seen this all over China on my frequent trips prior to my retirement.

The activity had increased measurably since my last trip. The Chinese were planning ahead for more future growth.

"Looks like Gary Hildegard is working really late," Terri said. "I just received an e-mail from him on my Blackberry about Dennis Dureno."

"Yes, he is," I replied. "It's one thirty in the morning in Beaufort. What did he say?"

"He apologized for taking so long to respond. He said things had been crazy there. It's tourist season, and there is never a dull moment during that time!

"Here is his response on Dennis. It seems Mr. Dureno is living completely on the edge. He has a big house on Lake Norman, two boys in university. One at Notre Dame and one at Princeton. His wife is a socialite involved in several charities and spends way above their income. Gary believes Dureno has everything he has invested in this product and the company going public. Best Gary can find out, he is currently living from credit card to credit card! He certainly cannot afford anything to go wrong, or he will be busted!"

"I'll be damned! I knew there was something going on with him. He was too pushy, almost in a panic about the product being shipped on schedule," I replied. "I'm not sure how he fits into the picture, but we need to find out more as we go along here."

"Do you think he had anything to do with Mr. Zhu's murder?" Terri asked.

"I'm really not sure. There is no doubt he has a motive to eliminate any obstacle that could affect production, but we don't have enough information on his whereabouts at the time of the murder. This something we do need to follow up on this week."

Terri sent a thank you e-mail back to Gary thanking him for his help and promising to keep him current on the details of the case.

"Gary did say in his e-mail he sure wished he was here to help us," Terri said. "He said he missed working with you and saving your butt!"

I sent him an e-mail from me. It said for him to be ready to jump in sometime in the near future. I might need his backup to save my butt one more time!

The train had reached a speed of 270 kilometers per hour, and the ride was smooth as silk. We made stops at Hunshan and Suzhou before arriving at Changzhou on time. We had three minutes to exit the train, so we were in line in the aisle before the train stopped at the station.

Changzhou station was one of the older train stations in China. A larger new one was in the early stages of completion across the tracks from the current one. The route to reach the exit door was filled with travelers entering and leaving the station, so we literally had to jostle our way to the exit door.

On the outside, the arrival and departure areas were situated next to each other, causing a massive traffic jam in front of the station. The mountain of people became jumbled puzzle pieces trying to fit themselves into arriving cars, taxis, and busses. The puzzle scenario was repeated for those attempting to leave the station as well. Police officers assigned to control the traffic had long ago given up trying to complete the puzzle. They stood by smoking, talking, and occasionally making a motion toward a car, taxi, or bus. People were shouting, trying to gain the attention of someone they were trying to pick up, trying to create a driving lane to move their vehicles forward, and at times shouting at each other, showing fists and using foul language!

Three station maintenance employees were attempting to wire several crowd control barricades together without much success. You could see the plan was to connect the temporary barriers to a permanent wall protruding from the train station. Even though there were signs forbidding traffic in the area, people continued to stream through the space between the temporary barrier and the wall, making their work almost impossible, but they continued to work diligently on their assignment, each one taking frequent smoke breaks, further complicating the process.

Mr. Da, owner of Universal China Production Company, had agreed to have one of his drivers meet us outside the station and drive us to the Grand Hotel and then to the factory for our 4:00 p.m. meeting. We found a small open spot on a street divider in front of the station and surveyed the ocean of oriental faces holding up signs with people's names displayed on them. We spotted a young man in a blue work uniform and hat waving a sign saying Mr. Rick and Ms. Terri.

We got his attention by waving to him as we made our way to his location. "Ni hao," I said.

"Ni hao. You Mr. Rick?" he asked.

"*Shi,*" I said.

He motioned for us to follow him as he pushed his way free of the immediate crowd and headed for the parking lot. The parking lot was a cross section of the rest of the outside area. Cars and vans were jammed together, some waiting for a spot to open up, while others made a vain attempt to circle the lot, looking for open parking spots.

Our driver pointed to a blue Buick van sitting in the third row of the lot. He placed our hand luggage in the back of the van and motioned us to the right side of the van where he opened the sliding door. I slid across to the left side of the first row of backseats. Terri took the seat near the window.

"To hotel Grand?" the driver asked.

"*Wei,*" I replied.

Our plan was to initially go to the Grand Hotel and check in, while hoping our bags would arrive about the same time. We would go to the plant after we had checked in and had our bags sent to our room.

Terri's cell rang as we exited the parking lot. "Hello? Hi, Gregory. Yes, we are in Changzhou. We are just now leaving the train station parking lot and heading for the Grand to check in and retrieve our luggage.

"How did your morning go?" Terri asked. "That's good to hear. Then you have an agreement with Mr. Xue for his firm to represent John? That's terrific. I'm sure Ann will be pleased and excited about the prospect of seeing John.

"No, we don't have any dinner plans," Terri said, looking at me for a signal. I nodded. "Yes, we can meet you for an early dinner at the hotel. Our meeting at the plant shouldn't last too long. What time is your train back to Shanghai? That should be late enough. We'll plan to meet you at six in the hotel dining room. Bye."

I said, "It looks as if everything went okay at the lawyer's."

"It sure sounds so. Gregory was very upbeat during our conversation."

"Had Mr. Xue received the data file from Ms. Chen?"

"Gregory didn't say. We can ask him that at dinner. Are you going to bring up what we heard from Gary about Dennis?" Terri asked.

"No, I don't think so. I want that one to simmer a bit before I say anything."

The Grand Hotel is the only five-star hotel in Changzhou and is situated in the center of the city on a very busy West Yangling Road. The front area of the hotel is miniscule in comparison to the size and location of the hotel. Our driver joined the queue of taxis and cars in the street waiting for an open position for us to unload. The area in front had a capacity of five vehicles at any given time; however, that number could have doubled if a row of parked cars facing the street had been eliminated.

The hotel's doormen were dressed in military-style dark gray suits with pillbox hats trimmed in gold. Each doorman had varying amounts of light gray braid on each sleeve, signifying his rank. A very frustrated doorman gave our driver a hurry-up wave when the next spot opened up.

Our driver leapt from the van almost before the van stopped to reclaim our hand luggage and open the sliding door for us to exit. There was a small piped lattice canopy overhang above the load and unload area with concrete columns supporting the lattice work on each side of the canopy. The columns were decorated with wrap-around banners of red and gold celebrating the new year.

Another doorman quickly greeted us and asked to carry our hand luggage. We declined, and before we walked into the main lobby, we made sure our driver would be waiting to take us to the plant after we checked in.

The center point of the main entry was a set of revolving doors of glass and gold with gold-trimmed normal entry doors on each side. All of the doors led into a mildly opulent entry with highly polished cream-colored marble floors with a small brown stripe pattern accented with an occasional oversized brown stripe.

The hotel itself consisted of twin towers with the ends of each building facing the street on a ten-degree angle. The seventeen-story right tower was wide with the lighted hotel name sitting atop an orange façade on the top floor and was known as the Emperor Tower. The Jade Tower was the second tower and was twenty-eight stories tall with a slender design. There were five buildings within the hotel complex that contained an expansive lobby, restaurants, shops, a workout club, and sitting areas.

The focal point of the lobby was a semicircular floor-to-ceiling mirror with a display of various flowers in clay pots occupying the floor with long-stemmed white flowers in sprays reaching approximately five feet high. A red lantern with yellow tassels hung from the center of the display. Speakers emitting Chinese music were positioned on wooden stands on each side of the display.

One of the Western restaurants was on the left of the display as you entered the lobby, and hotel registration was ninety degrees to the right. The marble lobby floor made a smooth transition from cream into full brown in the section around the registration desk.

We approached the desk and were greeted by a young clerk dressed in a black suit with white shirt and gold tie featuring a design of the hotel. His name tag said he was Roger.

"Mr. and Mrs. Watson, it's a pleasure to have you as our guests at the Grand. Your luggage has arrived and is stored in our secure storage area. I will be most happy to send it to your room after you have checked in."

I always marveled at how well all employees in major hotels spoke English. It was clear all major establishments had worked hard to make sure a Western guest was comfortable.

"I have taken the liberty to upgrade you and Mrs. Watson into a suite at the request of Mr. Da. He is a very good friend of the hotel! I hope that is acceptable?" he asked.

I replied, "Yes, that is more than acceptable. I will thank Mr. Da when I see him this afternoon."

"Very well! Your room number will be 5406, located in the west tower, close to the executive lounge. Is that acceptable?"

"Yes, it is," I replied.

We completed the check in process, thanked Roger, and walked outside. Our driver was standing next to one of the doormen, engrossed in conversation and smoking a cigarette as we exited the hotel. When he saw us, he immediately dropped his cigarette and motioned that he was going to retrieve his van.

He returned in about five minutes and patiently waited his turn to pull into the lobby drop area.

Mr. Da told me the drive time to his plant from the hotel would be approximately forty minutes at this time in the afternoon. He said it could be at least double that during evening and morning rush hours.

Changzhou had certainly changed since my last visit, and it was difficult to track our route to the plant, which was located in one of the industrial zones on the outskirts of the city. Some sights did remain familiar along the way. I recognized the Jian Lan Gearbox Company that I had purchased product from several years ago. Also, the Black Peong Company involved in the production of house slippers, and the Karl Mayer Company. Karl Mayer was a manufacturer of textile machinery used in the remaining South Carolina textile industry plants.

We passed Changzhou Engineering and Technology College as we neared the plant. I had coordinated closely with the college in the late nineties, developing curriculums for manufacturing personnel support.

United China Production Company (UCPC) was built in 1995 and typified many Chinese production facilities. The production buildings were mega-sized with high ceilings, marble floors, and open doors at each end of the building with very little heat for the winter in the plant. Offices normally were not heated during the winter months to conserve restrictive energy.

I remembered spending weeks in China visiting various suppliers' plants during the winter and wearing my winter coat during negotiation sessions held in conference rooms in the main offices. Many times, I survived by drinking substantial quantities of hot tea!

UCPC's footprint consisted of a four-story main office building and five manufacturing buildings containing a component production building, an assembly building, a heat treatment facility, an inventory building that held all the material used to produce the products, and two shipping and distribution buildings. Everything visible was painted white or blue. Each building was white inside and out with blue trim.

Contemporary Chinese businesspeople competed to see who could build the most grand office and manufacturing facility. An extremely large bronze-cast bull in full attack mode was the centerpiece of UCPC's main courtyard and entrance. During our time at UCPC, I asked several people why the bull was the symbol of the company and never received a definitive answer. Before Terri and I left UCPC for the last time, I planned to ask Mr. Da.

An expansive mural depicting lean manufacturing practices and encouraging each employee to comply hung in the center of the outside wall on the end of the main assembly building. It needed a face lift.

Two uniformed guards dressed in blue uniforms with resplendent gold braid over their shoulders stood waiting for a third in the guard house to open the accordion-style gate in the front of the plant. The guards saluted smartly as we passed by into the courtyard parking area.

Mr. Da Wei Huang and several of his staff greeted us with bows as we exited the van. Terri and I bowed in return and greeted Mr. Da. "Ni hao, Mr. Da."

"Ni hao, Mr. and Mrs. Watson. My pleasure to meet you."

"Ni hao, I am Jianping Xu. My English name Jan, and I am the export manager. President Da speak little English, so I translate for him."

"Xie xie, Ms. Xu," I replied to her as I was looking at Mr. Da.

"Xie xie, Mr. Da," I said, looking directly at him. "We are very pleased you have taken time from your very busy day to speak with us. I promise our visit will be short and meaningful."

Ms. Xu translated my English words to President Da in Chinese, "*Wo Men Gao Xing Nin Zai Bai Mang Zhi Zhong. Chou Kong He Wo Men Jian Mian. Wo Bao Zheng Wo Men De Lai Fang Jian Duan He You Yi Yi De.*"

Mr. Da answered, "*Wo Yi An Pai Hao Zheng Tian De Shi Jian Yi Bao Zheng Ni Men De Lai Fang Jiang Fu He Li Men De Qi Wang Bing Que Hu Hu.*" "*Yu Wd Men Shuang Fang. Quing Wo Lai.*"

Ms. Xu translated for Terri and me.

Mr. Da was at least six feet three and in immaculate physical condition. His suit was Armani and he wore a Rolex watch. No doubt he was a consummate believer in capitalism!

We followed him across the court yard and into the main office building. Stairs to the upper floors were directly in front of us as we entered the building. There was no elevator to the upper floors as in most Chinese factories and offices. Mr. Da led us down a hallway that ran along the outsider wall of the building. Several offices were positioned along the hallway on our left. We passed the offices of the Controller, manufacturing manager, purchasing, and human resources and entered a small conference room.

A long oblong mahogany conference table was centered in the room with four black heavily padded leather wheeled office chairs on each side. A large communications cabinet made of the same mahogany sat at the end of the table against the wall furthest from the door. The communications cabinet contained a chalkboard, flip charts, AV equipment, and a pull-down screen used for projections.

Mr. Da and Ms. Xu took the two middle seats on the right side of the table where their laptops were positioned. Terri and I took the opposing two seats. An office assistant entered the room with cups of Chinese tea for each of us. Mr. Da had a special cup with a different tea. I learned later it was a special blend made just for him and never shared. Ms. Xu asked if tea was acceptable, or is we would prefer coffee, juice, or Coke. We replied tea was just fine.

Once tea had been served, Mr. Da and Ms. Xu presented their business cards to us. Terri and I had developed a business card for each of us in anticipation of these kinds of situations. Our cards were printed on one side in English and the other side in Chinese. We held our cards in both hands, using our thumb and forefinger with the Chinese wording facing up, and bowed as we presented them.

Trading business cards is a very special act in China along with each individual's title. It is always important to read each card when it is presented and acknowledge receiving it. It is also proper to lay all the cards on the table in front of you for reference during any meeting. Business cards are always treated with respect.

I began, "Mr. Da, thank you for meeting with us and for arranging a suite for us at the Grand Hotel. Terri and I have a small gift for you, and for you as well, Ms. Xu.

"Mr. Da, I understand both you and your son are American National Football League fans, and your favorite team is the New England Patriots. I hope you will both enjoy these small gifts to thank you for meeting with us today."

I handed Mr. Da two packages, one for him and one for his nine-year-old son. I had placed the gift for his son in a gift bag so his father could examine it prior to giving it to his son. Mr. Da's gift was a regulation NFL football signed by Tom Brady. The gift for his son was a Patriots jersey with Tom Brady's name and number on the back.

Mr. Da opened his gift, smiled broadly, and thanked us profusely. There was no doubt our gift hit the mark. He inspected his son's gift and responded with another huge smile and thank you. "Son will be most pleased," Mr. Da replied in English.

Terri had selected a sweet grass basket purchased in the Charleston Market as a gift for Ms. Xu. Ms. Xu opened the gift and smiled. She said, "I am very familiar with the special nature of the sweet grass basket. I have seen them in travel magazines I have read, but never in person. This is a very special gift and will display in my home. Thank you for your kindness."

Mr. Da said, *"Gau Xie Ni Zai Ci Lai Fang." Wo Yan Yi Xiang Ni Jie Shao Wo Gong Si De Gai Kuang." Xu Xiao Jie Jiang Jie Jiang Fang Yi Ge Duan Piau.*

Ms. Xu translated for Mr. Da,: "Thank you again for coming here. I would like to give you an overview of my company."

Ms. Xu continued, "I have a short video about our company to show you, but before I start it, I would like to give you a few facts about United China Production Company. Our company was founded by Mr. Da's father in the year 2000 to produce high-quality machining, weldments, and fabrications mainly for companies like WMT. Mr. Da's father died in 2005, and Mr. Da became president of the company. Our yearly sales are $200 million US split between the United States, 50 percent; Europe, 20 percent; and Asia, 30 percent. We employ 800 people. May I begin?"

We said yes.

The video was informative and gave us a good idea of how the company grew to today's size, the plant layout, and quality process.

Mr. Da began when the video was over, "*Wo Gao Xing Ni Zai Zhe Lai Bang Zhu Que Ren Yi Fa Sheng De Shi Bing Shi Sheng Chan Hui Fu Dao Ji Hua Shang Lai. UCPC He WMT Yi Zai WMT De Chain Pin Shang Tou Zi Hen Duo Bing Qie XU Yao Jin Kuai Shou Hui Tou.*"

Ms. Xu translated, "I am pleased you are here to help determine what happened and to get production back on schedule. UCPC, along with WMT, has much invested in development of product and both need to realize investment as soon as possible."

"Mr. Da, I fully understand your concern regarding getting production back on schedule; however, my main reason for being here is to find who killed Chief Engineer Zhu Zhong Huang and help prove Mr. John Alworth's innocence."

Mr. Da replied after listening to Ms. Xu's translation, "*Wo Qi Wang Zhao Dao Shui Dui Wo De Hao Peng You Zhu De Si Fu Ze. UCPC Cong Lai Mei Fa Sheng Zhe Zong Shi. Ni You He Ji Hua Er wo Neng Ru He Bang Zhu Ne?*"

Ms. Xu translated Mr. Da's comments for Terri and me. "I wish to find who is responsible for the death of my good friend Zhu. Never something like this happen at UCPC. What plans do you have and how can I help?"

"Thank you, Mr. Da, for your offer to assist us. First, I would like to begin working in your plant tomorrow morning in the quality control area to review all the data regarding product testing. I am specifically interested in the additional life test Mr. Zhu ran."

Ms. Xu said in Chinese, "*Xie Xie Da Xian Sheng Zhu Dong Bang Zhu Wo Men. Shou Xian. Wo Xiang Ming Tian Zao Shang Zai Ni De Gong Chang Kai Shi Gong Zou Zai Pin Guan Bu Shen Cha You Guan Chan Pin Shi Yan De Sao You Shu Ju. Wo Te Bie Dui Zhu Xian Sheng Zuo De Shou Ming Shi Yan Gao Xing Qu.*"

Mr. Da looked at both Ms. Xu and me with surprise, not seeming to understand what I had said to him. He drank his tea and thought for a few moments before answering. He said in English, "What special test do you speak?"

There was no need to translate that. I fully understood what he was saying and I responded before Ms. Xu began to translate. "According to Mr. Alworth, Mr. Zhu ran an extended life test without permission, and those tests uncovered a fatal flaw in the design of the product. Mr. Alworth and Mr. Zhu had a very loud argument in the quality control lab regarding that test the morning of Mr. Zhu's death. Were you not aware of that?"

Ms. Xu slowly translated to Mr. Da: "*Ju A Er Wo Fu Xian Sheng Suo Yan, Zhu Xian Sheng Zai Wei Huo Pi Zhou Shi Zou Le Kuo Zhan Shou Ming Shi Yan, Er Shi Yan Shi Chu Chan Pin She Ji Zhong De Yi Ge She Ming Que Xian. A Er Wo fu Xian Sheng He Zhu Xian Sheng Si Wang Dang Tian Zao Shang Zai Pin Guan Shi Yan Shi Nei Guan Yu Shi Yan Fa Sheng Le Ji De Sheng Chao.*"

It was obvious Mr. Da knew nothing of the extra tests, the outcome of the test, nor the argument. He was stunned and embarrassed!

Mr. Da looked at Ms. Xu and said in a very loud and angry voice, "*Wo Te Sheng Xian Sheng Wo Yi Han Wo Bi Xu Jie Shu Zhe Ci Hui Jian Zhi Dao Ming Tian. Wo You Hen Duo Dong. Xi Yao Jin Tian Xiao Hua Quing Ming Zao Jiu Dian Zai Lai.*"

Ms. Xu quietly translated. "Mr. Watson, I am sorry I must end this meeting until tomorrow. I have much to learn today! Please return tomorrow at nine a.m."

With that, Mr. Da shook our hands and quickly left the conference room.

I would learn later that Mr. Da ran a very tight operation and prided himself on his knowledge of the day-to-day operation of his business. He did not tolerate being embarrassed! I was glad I was not an employee of UCPC!

Ms. Xu apologized profusely for the abrupt end to the meeting and suggested their driver pick us up at our hotel at 8:00 a.m. the next morning. We agreed.

Our driver was standing next to the car, waiting to take us back to the hotel as we exited the building. We thanked Ms. Xu for her assistance. Once again, she apologized for Mr. Da. We accepted one more time as we entered the backseat of the van. Ms. Xu gave the driver our destination, and we exited the courtyard of the plant to the salutes of the guards at the main gate.

We arrived back at the hotel around 6:00 that evening with just enough time to drop our briefcases and computers in our room and freshen up for dinner.

As promised, our luggage was in, placed in the bedroom of our suite, when we arrived. The room was spacious with a living room, bedroom, conference area, and two bathrooms. Roger had chosen a Western-style suite that featured Ansell Adams photography and a black and white theme for the furniture and bedroom.

The furniture in the suite was modern in design with high gloss black finishes throughout. A long black couch sat along the wall directly opposite the entry door with a loveseat to match. The end tables and coffee tables were high gloss white with black swirls flowing across their tops. A small conference table, also high gloss black, with four white leather chairs with black high gloss wood trim completed the suite's living area.

The bedroom's centerpiece was a king-sized bed with a black high gloss headboard decorated with white swirls to complement the end tables and coffee tables in the living room. The bedspread was solid black, and the lamps were white.

The suite showcased several Ansell Adams photographs of the American East Coast.

It did not take long to unpack our bags and get organized for what we hoped would not be a prolonged stay.

Gregory had made reservations at the Galaxy Restaurant located on the top floor of the Jade Tower. The Galaxy was the top-rated restaurant within the hotel's twenty-six eating establishments and was well known in Changzhou as the city's best Western restaurant.

Gregory was seated when we arrived. He saw us enter and waved us to his table. The Galaxy would become Changzhou's Big Bamboo. Terri and I would try and enjoy a wide variety of exquisite Chinese restaurants both within the hotel and around the city during our stay, but the Galaxy became Gregory's favorite and our meeting place to discuss the day's events with Gregory, Ann, and Dennis.

The Galaxy reminded me of Carolinas, one of our Charleston favorites. The décor was comfortable with both booths and tables. We always requested a table at Carolinas in the front of the restaurant that looked out to a Civil War cobblestone gas lamp—lit street. Gregory had selected a window table in the Galaxy that had a commanding view of the city below and across the horizon. He always preferred restaurants with views that were situated high in a building. I always believed it fed his ego regarding what he believed was his place in the world. Always taking the high ground, leading and being envied by everyone.

"Hey, guys, good to see you. Glad you could make it," he said. The waiter seated us across from him in his booth. "I hope your day was better than mine!"

"Why?" I asked. "I thought you told Terri your meeting with Changzhou, Huaide, and Zhanglinfang went well?"

"Oh, the meeting went okay, and Mr. Xue was gracious enough, but events are moving quicker than I would like. I always expect to be in control of events, but so far, that's not the case with everything surrounding John and the murder."

He continued, "The Changzhou police have transferred John here as scheduled, and Chief Inspector Ji of the Changzhou police has already visited UCPC and has gathered testimony from several so-called witnesses that were in the plant on the day of the murder."

"Mr. Xue has appointed Sun Jianguo to John's case. He is a senior member of the firm with a strong history in high-profile cases and is familiar with manufacturing. He met with the chief superintendent of police here in Changzhou early Monday morning in anticipation of our reaching an agreement. said

Terri asked, "Did you meet with Mr. Jianguo?"

"Yes, I did, and I like him, especially his criminal and manufacturing knowledge as well as his familiarity with dealing with the Western world."

"Did Mr. Jianguo say what happened when he met with the chief superintendent?" Terri asked.

"He told me the Changzhou police were under great pressure to get this case solved. Mr. Da's father was a high-ranking party official and was well respected in high political circles. The police want a quick conviction and a public appreciation for their work."

The waiter had placed menus at our table, but our conversation precluded any immediate review and selections. We decided to take a break from our conversation and order after our waiter made several futile passes by our table.

Gregory ordered a T-bone steak rare, and Terri and I ordered grilled chicken. Gregory asked Terri to select the wine. The Galaxy's wine cellar was extensive, making her decision somewhat difficult. She finally settled on a Cakebread chardonnay and a Ridge zinfandel, both produced in California.

"I am sure Mr. Da did not know Chief Inspector Ji had been in his plant," I said. "In fact, he was very upset when I told him about the extended life tests run by Mr. Zhu and angrily cut our meeting short this afternoon. He left the meeting on the warpath to find out who, when, and why. I told Terri, I really was glad today I did not work for UCPC, especially anywhere around the quality lab!"

I continued, "I have the impression Mr. Da runs an extremely tight ship and just couldn't believe anything like all of this could happen in his plant without his knowledge. He will not be happy when he finds out Chief Inspector Ji has been in his plant and interviewed his employees without his knowledge! I feel comfortable the Changzhou chief superintendent of police received an earful by the end of the day!

"I fully expect we will meet with a totally different Mr. Da tomorrow morning. I am sure he will be fully informed of everything by then!" I said.

Terri asked, "What about John's bail? Any news on that?"

"As of today, there will be no bail for John," Gregory said. "Mr. Xue and Mr. Jianguo are unable to request bail until John has been formally charged, and the Changzhou police are not ready, as of today, to do that. I believe they are gathering their information and possible evidence in preparation for bringing a formal charge, but are not ready today."

"Any idea when that might happen?" I asked.

"Not sure. Mr. Jianguo would like to meet with both you and Terri tomorrow morning to discuss what you know about the case and how you can work closely together. Then, I believe he will again meet with Chief Inspector Ji. When is your next meeting with Mr. Da?"

"He asked us to return at nine in the morning, but I am sure we can reschedule with a call to him early tomorrow morning," I answered. "I will request our meeting be moved into the afternoon so we can meet with Mr. Jianguo around mid-morning. I assume you are still taking the train back to Shanghai tonight?"

"Yes, but I plan on returning Thursday morning. I hope we will have enough data by then so I can make a decision on whether or not to begin production. As I have said, I am fully committed to proving John's innocence, but further delays in production could create unrest with our backers and the possibility of missing our first delivery commitments. Any of those events could be fatal to our business. My fervent hope is that the murder and producing product are independent of one another."

"That would be the best of both worlds, of course," I replied. "I just hope we will be far enough along in our investigation to assist you in making that decision."

We continued to discuss the day's events throughout dinner and finished with just enough time for Gregory to make his 10:00 p.m. train. We agreed to call Mr. Jianguo early Tuesday morning and arrange a meeting time compatible with our meeting with Mr. Da, and then talk with Gregory early Wednesday evening prior to his Thursday return trip to Changzhou.

Terri and I had a nightcap in the lobby bar and returned to our suite. We spent the next hour discussing our plan for Wednesday's meetings with Mr. Jianguo and Mr. Da. I wanted to finish our meeting with Mr. Da as soon as possible so I could begin my work in the quality lab. Terri would use Ms. Xu as her interrupter and interview as many of the same people that Chief Inspector Jihad as she could. I would locate Mr. Zhou's assistant who, according to John, spoke excellent English, and begin to review all the data on the product, looking for any clues to what happened during the additional extended life test on sample eight. My first order of business would be to find that original data.

We were both ready for bed by 10:30 that evening and set our alarm for 6:00 in the morning, leaving time for stretching and running at least eight miles followed by steam, shower, and breakfast.

The beds were so comfortable both of us really did not want to get up when the alarm went off at 6:00 a.m. It took several minutes of pestering Terri before she made it out of bed, brushed her teeth, and changed into her running gear, ready to begin the day. I beat her and was sitting on the bed waiting when she came out of the bathroom. "God, why are you such a pain in the ass every morning? Can't you just once be late or grumpy? No, don't answer that. I really don't want to know!" she said as she bolted for the door, leaving me sitting on the bed. "Come on, let's go. What, do you think we have all day?"

"Now who's the pain in the ass?" I yelled as she went out the door.

Chapter 13

We began Tuesday, May 20, with our normal workout, steam, and a small breakfast of fruit and cereal by 8:00 and headed to our suite to shower and call Mr. Jianguo.

The receptionist at Changzhou, Huaide, and Zhanglinfang placed me on hold while she paged him. He was in the building, but not in his office. I waited only a minute when I heard a click and Mr. Jianguo said, "Ni hao."

"Ni hao, Mr. Jianguo," I said. "This is Rick Watson, *Ni hui shuo Yingyw ma?*"

"Oh, yes, Mr. Watson. I have been expecting your call, and as you can see, I do speak English."

"I apologize for my ignorance! Of course you would speak English. Please forgive me."

"Nothing to forgive, Mr. Watson. A common mistake," Mr. Jianguo said. He was laughing as he responded. "I look forward to our meeting today. Do you know when you can be here?"

"Would ten this morning be convenient?" I asked.

"Ten would be fine. Would you like for me to send a car for you?"

"Thank you, but that won't be necessary," I replied. "We are staying just a few blocks from your office at the Grand Hotel, so we plan on walking."

"Excellent. Enjoy your walk, and I will see you at ten," Mr. Jianguo said.

Terri called Ann, hoping we could talk to her before our meeting with Mr. Jianguo at 10:00, but her cell went to voice mail. The message asked Ann to return the call. Terri and I wanted to know when Ann would be in Changzhou so we could arrange to see her.

The office of Changzhou, Huaide, and Zhanglinfang was three blocks from the hotel. A right out of the hotel, two blocks down, and one block to the right. It was a hot and humid morning where the smog seems to fill every pore of your skin, making each step an adventure.

"Tell me again whose idea this was?" Terri asked. She was perspiring as if she had just completed a morning run. "If I find the person who said, 'No, we don't need a car. We will just walk,' it will not be pretty! Look at me, I'm already a mess!"

"Oh, come on, it's not that bad, and it's only three city blocks. Besides, you always look terrific."

"Too late now, buster, you are in the dog house big time!"

The office was located in the middle of the block. It was a very old traditional Chinese building among modern high rises. The building was built in the pagoda style with rich reds, blues, greens, and a yellow tile roof. Each peak of the roof was filled with lions of varying sizes, depending on the distance of the lion from the building wall. The lions along with mythical animals were placed there as guards to peace and happiness. I wondered what had to happen to save this building in one of the high-rent districts in the city. In the United States, a building like this one would be declared a national monument, but only after years of lobbying and the placement of strategic campaign contributions.

We entered through one of the two ornate, heavy, wooden, steel-trimmed entrance doors into an elaborate lobby decorated with equally intriguing art and sculptures. We were surprised to see Ann Alworth sitting uncomfortably in a straight wing-backed chair with dragon designs throughout.

I don't remember who actually spoke first, but I remember Terri saying, "Well, Ann, this is a nice surprise. I left you a voice mail on your cell early this morning. When did you arrive in Changzhou, and what are you doing here?"

"I took the early morning train from Shanghai. I talked to Gregory late last night, and he told me John had already been transferred here and that Mr. Jianguo was going to be our lawyer. I called Mr. Jianguo from the train to arrange a time to meet with him, and he informed me he was meeting with you at ten, so here I am."

"That's great," I said. "Now we can all discuss our next steps and set a plan in motion to prove John had nothing to do with the murder. I don't suppose you have arranged a time to see John, have you?"

"No, I thought that was something we could discuss this morning," Ann said. "Frankly, I'm getting a little frustrated with everything surrounding this. All I want to do is see my husband and help you get him safely back home."

"And I guess we can add, get the product released to production as well," I said. "I know proving John had nothing to do with the murder is paramount, but the initial production of the product just can't seem to be separated even though I want it to be."

Mr. Jianguo's assistant approached us as we were finishing our conversation.

"Are you Mrs. Alworth and Mr. and Mrs. Watson?"

"Yes, that's us," I replied.

"Please follow me. Mr. Jianguo is expecting you."

The elevator took us to the third floor and opened in the center where the offices spread both left and right. The floor housed a particular group of councilors, each one tied together by expertise and case loads.

Mr. Jianguo's office was situated in the far left corner of the building. We made our way through several aisles separating offices of support staff to a very distinguished group of individual lawyer's offices on the parameter of the floor.

Mr. Jianguo was waiting for us in his office door and spoke first to Ann upon our arrival. "You must be Mrs. Alworth. I am Mr. Sun Jianguo, and I am your husband's counsel."

Ann replied, "Yes, I am Mrs. Alworth."

Mr. Jianguo turned to toward Terri and me, taking Terri's hand, and said, "And you must be Mrs. Watson, and you"—pointing to me—"Mr. Watson."

"Yes, we are, Mr. Jianguo, and I must say, it is a pleasure to meet you. Our friend and CEO of WMT, Mr. Gregory Brightson, has told us, in his opinion, you are the one to help us with this case."

"He is most kind. I must work hard to meet his and your expectations. Now, please enter, and let's discuss the case and how I can assist you and your husband."

Mr. Jianguo's office was at least equal to Ms. Chen's office in Shanghai. Large full-length windows overlooked the city, and the furniture was made of leather and very comfortable. The office contained a large desk as the focal point of the room. It was a light mahogany with ornate Chinese designs carved through the whole of the desk. The conference table was of the same design. We learned later each section of the table top and desk top related a story concerning the various areas of China.

Mr. Jianguo pointed to the conference table and led the way there. Ann, Terri, and I took seats on one side of the table and Mr. Jianguo the other.

Mr. Jianguo led off the conversation. "First, we will be working very closely together on this case, so please call me Sun. May I call you Ann, Terri, and Rick?"

I said, "Yes, please do call Terri and me by our first names, and you are correct; we certainly will be working closely together on this case, which I hope we can wrap up as quickly as possible." Ann nodded her agreement.

"I have read the notes from Ms. Chen regarding this case and have talked preliminarily to Chief Inspector Ji to ascertain what evidence the police have that implicates Mr. Alworth. Everything they have is circumstantial except Mr. Alworth's gun, which was found at the scene. Many people heard the argument between Mr. Alworth and Mr. Zhu, but no one heard the fatal shot nor saw anything out of the ordinary at the time of the shooting.

"This afternoon I plan to visit Mr. Alworth and have my first talk with him. I will also be requesting bond so we can remove Mr. Alworth from jail," Mr. Jianguo said. "Mrs. Alworth, are you available this afternoon to accompany me on my visit to the Changzhou police? I apologize for the short notice."

Ann replied, "Definitely, I will be there any time you wish, Mr. Jianguo. I am desperate to see my husband. It has been long enough, and I am really losing patience."

"I fully understand your frustration; however, our police force sometimes has little regard for any family relationships in cases like these. In your husband's case, the Changzhou police are confident he is the killer and that it will just be a matter of time before all the evidence will come forward to convict him.

"Now, Mr. Watson, er, Rick, can you please inform me of your and Terri's involvement in this case. I have some information from Mr. Brightson, but would like to hear it from you."

I began, "This may be a little hard to explain, but I will try. John, Terri, and I have been both professional and personal friends for more than fifteen years. The day after he was arrested, he was allowed to make one telephone call. At that time, he did not trust his company to do what would be necessary to extract him from this situation, in which John says he is completely innocent, so he called me and asked that I, along with Terri, come to China to help him prove his innocence."

"And according to Mr. Brightson, you also have had experience in law enforcement as well. Am I correct?" Mr. Jianguo said.

"Yes, I was in the military police when I was in the United States Marines and worked many criminal cases during my time in the service," I replied.

"Thank you, Rick," he said. "I am now more comfortable with your involvement in this case. I am hopeful we can develop an approach that will fit both our needs and jointly find a way to prove Mr. Alworth's innocence."

Mr. Jianguo continued, "So, this afternoon, Mrs. Alworth and I will visit her husband, and I will attempt to establish the bond requirement. Tomorrow, I plan to be at the plant to interview as many employees of UCPC and the president and his staff to see what they know about the murder and who may have done it. It is my understanding that Chief Inspector Ji has only made one visit to the plant as well. May I ask what are your plans, Rick?"

"Terri and I have a meeting with Mr. Da this afternoon. We had an initial meeting yesterday; however, Mr. Da was not fully aware of what had happened in his plant, so he requested we postpone our meeting until this afternoon. Tomorrow, with Mr. Da's permission, I will begin in the QC lab looking for clues and any data I can find on this unauthorized life test and try to determine the results of the tests and how the results affect the product. Terri had planned, like you, on interviewing employees in the plant that work near the QC lab and anyone who may have something to do with the quality of the product."

Mr. Jianguo said, "Excellent. If Mr. Da does refuse to allow you or Terri in the plant, I can have the proper entry paperwork processed in a few hours. Terri, would you like to work with me?"

"Sun, working with you would be my pleasure."

"Fine. Why don't you and I spend some time this morning detailing the questions we would like to ask. I believe we can use a basic set of questions and make adjustments as we go along. You agree?"

"That is fine with me. When would you like to start?" Terri asked.

"One of the things I love about working with Americans is your immediacy. You quickly get right to the point. In China, it is much longer to get there," he said, laughing. "We can begin now."

We spent the next hour developing a list of questions we felt were critical to the case and discussed how to approach each person in the best way to obtain information. Mr. Jianguo suggested Terri work with him in some of the first interviews to get a feel for the questioning process.

It was approaching noon, and Terri and I needed a quick lunch in order to make our 2:00 in the afternoon meeting with Mr. Da. Sun suggested a restaurant close to the office. We agreed and proceeded to the restaurant which was, as Sun suggested, five minutes away.

The Fugi was located on a corner of two very busy streets and was obviously a popular noontime eating establishment. The lobby floor was cream-colored marble and led to a wide set of brown marble stairs to the second floor and the first floor of the restaurant. The lobby was decorated with red Chinese lanterns hanging in the ceiling, a pole covered with shiny red paper balls that stood at least seven feet tall; a small statue of a Chinese hostess dressed in traditional red was bowing and welcoming us. There were smaller Chinese female characters on each step up to the second floor.

The second floor landing also had a bowing hostess statue wearing traditional pink dress and a real one to greet us. We followed her to a private room. Sun was a frequent visitor and had placed our order during our walk to the restaurant. The initial dishes were on the lazy Susan waiting for us. We began with a cold group of appetizers, a selection of hot chicken, pork, and vegetable dishes, a whole fish, and the main course of either rice of noodles. Fortunately, the watermelon was delivered next, signaling the end of the meal. We were more than ready to end the exquisite lunch as our time to depart for Mr. Da's plant was near.

Mr. Da's driver was waiting for us as we approached the offices of Changzhou, Huaide, and Zhanglinfang. Once again, we thanked Sun for a fine lunch and promised to call after our meeting with Mr. Da. Ann said goodbye as agreed to meet us in the lobby of the Grand at 7:00 that evening for dinner. She would accompany Sun to the Changzhou police office later to have her initial meeting with John.

Our driver manipulated city traffic and managed to arrive at UCPC exactly at 2:00. Ms. Xu was standing in front of the main office building and greeted us as we exited the car.

"Ni hao, Mr. and Mrs. Watson." She was bowing slightly. "Welcome back to our company. It is a pleasure to see you again."

"Ni hao and xie xie, Ms. Xu," I replied. Terri and I returned her bow. "We are pleased to be here and to talk once again with you and Mr. Da."

"Yes, he is waiting for you in his conference room." Ms. Xu led the way to the same conference room we were in yesterday. Mr. Da was already in the room and greeted us with firm handshakes.

"Welcome back to my company. Most pleased you return."

"I also have gift for you and your wife." Mr. Da handed Terri a very carefully wrapped box with delicate Chinese wrapping paper. Terri understood that this wrapping paper was not to be torn, but unwrapped very carefully and saved.

The box contained a uniquely constructed Chinese fan with an original painting on the front depicting a Chinese philosopher sitting on a large rock with an evergreen tree behind him. He was sitting by a flowing river surrounded by mountains and colorful birds perched in trees and flying across the painting. There was an original poem on the back of the fan describing the scene and the philosopher's contemplations as he reflected on what he saw.

Ms. Xu explained to us the poem was dedicated to Terri and me and signed by the artist.

"We thank you for your very kind and thoughtful gift. We will display it in a prominent place in our home for all our guests to see and to always remind us of your thoughtfulness," Terri said to Mr. Da. I thanked him as well.

Ms. Xu translated our comments to Mr. Da, and he replied he hoped his gift was special. We agreed it was.

"Today, we have much to discuss," he said in English. Mr. Da pointed to two chairs on the opposite side of the conference table, while Ms. Xu asked what we would like to drink besides tea. Terri said the tea would be fine, but I asked for a Diet Pepsi. That request almost always throws any Chinese business host. China is primarily a Coke country. The middle of the table was filled with candy, small Chinese cakes, cookies, and fruit.

Terri's tea arrived and, believe it or not, so did my Diet Pepsi. When we were comfortably seated, Mr. Da began. "I apologize for ending meeting quick yesterday. Today I better informed." With that, he looked at Ms. Xu.

"Mr. Da spent yesterday investigating the items you discussed yesterday and is much more informed and ready to talk about them," Ms. Xu said.

Mr. Da began, "*Wo He Wo De Chang Zhang Tan Guo Bing Fa Xian Gui Fang A Er Wo. Fu Xian Sheng Zai Chang Bing He zhu Xian Sheng Jian Guo Ji Lie De Zheng Chao Er Qie A Er Wo Fu Xian Sheng Xiang Zhu Xian Sheng Da Sheng Pao Xiao. Wo Liao Jie Zhu Xian Sheng Dui Chan Ping Jin Xing Le*

Fu Jia De Shou Ming Shi Yan Er A Er wo Fu Xian Sheng Bu Xi Huan Ta Zhe Yang Zuo. Me You Ren Ting Dao Huo Kan Jian Qiang Ji. Yi Ge Wan Ban Qing Jie."

Ms. Xu translated, "I talked to plant manager and found out that your Mr. Alworth was here, had meeting with Mr. Zhu. I believe they had very bad argument, and Mr. Alworth was yelling loud at Mr. Zhu. Mr. Da understands Mr. Zhu ran additional life tests on product, and Mr. Alworth did not like it. No one in plant heard or saw gun being shot. One of our cleaning women found Mr. Zhu dead on floor in quality control office."

Mr. Da said, "*Wo Wen Quin Jian Xin Fu Jai Shi Yan De Shi Yan Bao Gao Zai Na Li, Dan Shi Ta Bu Zhi Dao.*"

Ms. Xu said, "I asked Quin Jian Xin where test reports were of additional test, but he knew nothing of them."

I asked Mr. Da, "Did you know Chief Inspector Ji was in your plant interviewing your employees about the murder?"

Ms. Xu translated, "*Ni Zhi Dao Ji Tan Zhang Zai Gong Chang Nei Xiang Wo Men De Yuan Gong Xun Wen Guo You Guan Mou Sha De Shima?*"

Mr. Da responded, "*Wo Xian Zai Liao Jie Xi Xian Sheng De Dao Fang Bing que Bo Chu Fei Wo Yu Xian Zhi Xiao Ta Jiang Lai Bu Hui Jin Ru Wo De Gong Chang.*"

"I am now fully aware of Mr. Ji's visit and have made sure he will not enter my plant in future unless I know first," Ms. Xu translated.

"Did the chief inspector tell you what he found while he was here?" I asked.

Mr. Da understood what I asked and answered in English, "No, he not told me anything of visit, but I call chief superintendant for information."

"Did the chief superintendant respond to you yet?" I asked.

Mr. Da answered, "No."

"May Terri and I begin our investigation in your plant today?" I asked.

Ms. Xu said to Mr. Da, "*Wo Te Sheng Xian Sheng He Fu Ren Ke Fou Jin Tian Zai Chang Nei Kai Shi Diao cha?*"

Mr. Da replied in English, "Yes, investigate."

"I would like to begin by talking to the plant manager," I said. "And Terri and Mr. Jianguo of Changzhou, Huaide, and Zhanglinfang would like to begin talking to your employees as well. Is that okay?"

Ms. Xu translated, "*Wo Te Sheng Xian Sheng Xiang Kai Shi He Chang Zhang Tan Er Wo Te Sheng Fu Ren He Jian Guo Xian Sheng Ye Kai Shi He De Yuan Gong Jiao Tan. Ke Yi Ma?*"

Mr. Da replied, "*Shi, Mei Wen Ti. Qing He Ni Xiang Jiao Liu De Ren He Ren Tan Hua. Dan You Bu Hao Di Xiao Xi. Chang Shang Bu Zai Zhe gong Zuo Le.*"

Ms. Xu said, "Yes, Mr. Da says no problem to talk to anyone; however, plant manager not work here anymore."

I looked at Terri and said quietly, so she only could hear me, "I guess he is just what everybody said, not one to mess with!"

I asked Ms. Xu if she knew where the plant manager might be if we needed to talk to him, and she said maybe she could locate him if we needed. She did not speak very confidently, and I was sure she would not like her boss to know about any contact she may have with the disposed plant manager.

"Ms. Xu, please ask Mr. Da if he would like to talk to his people before Terri and I go into the plant and speak to them," I said.

Ms. Xu replied, "Mr. Da already talk to everybody. Okay for you to go to plant now."

We thanked Mr. Da for his kindness and promised to meet again with him and Ms. Xu prior to our leaving for the day.

Ms. Xu communicated that to Mr. Da. He smiled and pointed to the door as a signal that the meeting was at a temporary end.

Ms. Xu lead the way outside the main office and to the main entrance to the manufacturing complex of UCPC. I gave the bull a quick pat for luck as we passed by, and Terri gave me her "are you kidding" look.

The quality control lab was located in the last manufacturing building on the left hand side of the compound. This building was primarily used for receiving goods, checking and processing raw and finished materials required for production of the main products produced by UCPC. Receiving inspection and the quality control lab were the hub of all activity in the building. Samples of every item received were required to pass through both operations and be accepted by quality control. Anything not accepted was placed in a quarantine area for further disposition prior to being released for use in manufacturing or returned to the offending supplier.

Mr. Zhu had worn two hats at UCPC. He had been responsible for both engineering and quality control. This was very unusual for Chinese companies, who normally split these important positions; however, Mr.

Zhu had been a part of UCPC since the company's inception and had gained Mr. Da's father's and now Mr. Da's complete confidence to manage both positions.

There was a very small assembly line that produced a low-volume specialized alarm used in steel plants. The assembly line was located there so the chief engineer could personally oversee the quality of the production for a very influential customer. Terri and Sun would begin their questioning of the workers on the assembly line first, because of the proximity to the QC lab office. I would begin in the lab itself.

Sun had arrived and was waiting for us in the QC lab. "Been here long?" I asked Sun.

"No, I have only been here for about fifteen minutes. Mr. Qian has been kind enough to keep me company while I was waiting."

"How did the meeting with the chief inspector go?" I asked.

"We did not get as far as I wished. Maybe we should have dinner at your hotel tonight with Mrs. Alworth and discuss it."

"Let's do that," I answered. "We can finalize the details after we complete the day here.

"Ms. Xu, this is Mr. Sun Jianguo from Changzhou, Huaide, and Zhanglinfang," I said. "His company has been retained to represent Mr. Alworth. He has also agreed to work with Terri and me."

Sun bowed to Ms. Xu and introduced himself in Chinese. Ms. Xu acknowledged and told him she would accompany him and Terri as they did their interviews of the plant employees.

Ms. Xu introduced Terri and me to Qian Jian Xin, the assistant quality control manager.

"Ni hao, Mr. and Mrs. Watson. It is very nice to meet you," Xin said.

"Oh, do you speak English, Mr. Xin?" I asked. I already knew he did based on my conversation with John in the Shanghai jail, but wanted to see his response.

"Shi shi, I do speak English. Mr. Zhu required it for my position. I always work with our customers on everything relating to the quality of their products."

"That's excellent," I said. "It should certainly make the investigation go more smoothly," I said.

Terri said, "Rick, Sun, Ms. Xu, and I are going to walk over to the assembly line and begin interviewing those employees while you begin here with Mr. Xin."

"Okay by me. Mr. Xin, how about you and I get started."

"Please, Mr. Watson, call me by my English name, Billy."

"Billy it is, and please, call me Rick. Now let's get started."

I spent the next two hours familiarizing myself with the equipment in the lab, the computer system, and the particular electronic and manual filing systems Mr. Zhu had developed to track each piece of raw and component material as it was processed through the quality control system. I also did a quick review of the UCPC quality manual. One of the processes described in the manual was the process to be used to qualify all customers' new products. With Billy's help, I identified each relevant section of the quality manual down to the detail level available that was used to test components for WMT's new product and the process used to perform life tests. I was immediately impressed with the ability to gather data and the data's thoroughness.

My next step was to compile a data flow chart using the quality manual as a reference. The flow chart outlined the sequence of each individual piece of information, assigned an identification number to the data for later easy reference, and ranked the importance of each data point along the flow chart. By charting the entire process that lead to the final product approval, I hoped to quickly isolate the additional extended life test that Mr. Zhu had run without authorization. It became apparent this task was going to take several hours, so we decided to save our data in my laptop and begin again tomorrow morning.

Terri and Sun were finishing interviewing for the day about the same time. We followed Ms. Xu to the main office building and the conference room where Mr. Da was waiting for our review as promised.

A young assistant placed hot tea at each seat and asked if I wanted Diet Pepsi. I said yes. She nodded and left the room just as Sun greeted Mr. Da in Chinese. I understood enough to understand it was obvious they knew each other prior to this case, as they exchanged pleasantries about families and Sun remembered Mr. Da's father.

Sun began in English and then translated for Mr. Da. "We are very grateful to you for allowing us to speak with your employees and work in the quality control area regarding this most unfortunate incident. I personally assure you we will do our best not to disrupt your daily operation and shipments to your most valued customers."

Mr. Da nodded and smiled at Sun's comments. He replied, and Sun translated for Terri and me: "I am very saddened with the death of Mr. Zhu and the accusation against Mr. Alworth. Mr. Zhu was good friend and loyal employee, and Mr. Alworth always respected my business and people. Your father and mine would not understood how such act could happen in our country today. Has life lost so much value?"

"Yes, I believe unfortunately it has, and your father and mine would not have tolerated such a thing; however, we are not sure if the murder was committed by a Chinese person or a foreigner," Sun translated.

Mr. Da. responded in English, "We must seek truth."

"We shall," said Sun.

"Mr. Da," I said, and Sun translated, "I wish to compliment you on the manner in which you control quality in your company. Today, I have examined your process, computer system, and filing. It is among the best I have seen."

Mr. Da again smiled and replied to Sun's translation, "I am humbled by your words."

"Today, Billy and I began a flow chart and data accumulation in an attempt to understand what happened in the approval of the WMT new product. I will return tomorrow, with your approval, and continue to unravel the puzzle." I looked at Sun and waited for his translation.

Mr. Da listened and replied in English, "Please, return and learn."

Sun also told Mr. Da what he and Terri had accomplished during the interviews and suggested a need to digest the data prior to further interviews. He also told Mr. Da that he was meeting again with Chief Inspector Ji tomorrow and hoped to have more details regarding the murder.

We thanked Mr. Da and Ms. Xu again and left the conference room. Our driver was waiting to return us to the Grand Hotel. Mr. Da and Ms. Xu waved goodbye as we left the compound.

Sun's driver was also waiting in the courtyard. Sun suggested Ann accompany him to the hotel so they could discuss tomorrow's meeting at police headquarters.

The traffic Mr. Da mentioned in our first conversation with him was out in full force on the way back to the Grand. Our two cars were quickly separated in traffic. The normal twenty-minute ride escalated to one hour and fifteen minutes, leaving Terri and me time to review what happened that day and do some planning for Wednesday. We arrived at the hotel ahead of Sun and Ann, giving us just enough time to do a quick refresh and meet Ann and Sun in the Galaxy for dinner.

We were the first to arrive and selected a table overlooking the city just as we knew Gregory would do if he were with us. Sun and Ann exited the elevator together within fifteen minutes of our arrival. We had pre-warned the head waiter of their pending arrival. He met them at the restaurant entrance and led them to our table.

"Good evening to you all. I hope we haven't kept you waiting. Didn't you all get stopped because of the accident just a few miles from the plant?" Ann asked. "We had to sit and wait. There was no way around it."

"No, I guess we were lucky," I replied. "Our driver was talking to someone on his cell about what I know now was an accident. We must have just missed it. Anyway, we are all here now, and thankfully, it did not involve us, so please join us."

"I don't know about you two, but I am ready for a good glass of wine and dinner," Terri said. "Mr. Sun, I assume you have eaten here before and a Western restaurant is acceptable?"

"Yes, I have eaten here many times, and I do truly enjoy a well-prepared Western dinner on occasion."

"Well," Ann said, "If you are going to hang around with us and especially Gregory Brightson, you will become very familiar with the menu here!" She laughed and pointed at Terri and me for confirmation.

Terri replied, "Ann is certainly correct. Rick and I have to sneak off if we want anything but Western food when Gregory is here."

We laughed at the situation and all agreed to coerce Gregory into exploring a gourmet Chinese restaurant during his next visit to Changzhou. Sun volunteered to select the venue and invited Mr. Xue to join us. Our waiter waited patiently until we had completed our conversation before describing the evening's specials to us. Either we were famished or he was magnificent in his descriptions of the specials because each of us ordered something different. Sun couldn't resist the T-bone steak, Ann selected the roasted breast of duckling, Terri decided she had an appetite for pork and ordered the pan-fried pork chop, while I found the old southern favorite, buttermilk fried chicken, enticing.

Once again, Terri made excellent California wine choices. She selected a 2005 Amber Knolls Sauvignon Blanc for the pork chop and the fried chicken, and a 2006 Santa Rita Hills pinot noir for the duck and the T-bone.

Our dinners were as advertised, and we leisurely moved through the dinner courses while discussing the current day's events and our plans for tomorrow.

"How did the interviews with the employees go?" I asked both Terri and Sun. "Were you able to determine anything about what happened on the evening Zhou was shot?"

Sun began, "We interviewed each person working on the assembly line, but no one could tell us any more than we already knew. Unfortunately, the two witnesses and the cleaning woman were not in the plant yesterday, but should be there tomorrow."

"Anything to the situation that all three of the individuals that have firsthand knowledge of the murder were not in the plant?" I asked.

"We originally were suspicious of that, but found out by talking to the line supervisor the two specific workers who said they witnessed the murder had been assigned to work in another of the UCPC plants for the day, and the cleaning woman was at the company doctor's office also late in the day," Terri said. "We will talk to them on Thursday."

Sun asked, "What did you find in the QC lab, Rick?"

"First, I am confident that whatever Zhou did was thorough. All of the data was meticulously kept and in fine working order; however, I did not find anything vaguely resembling any additional extended life test on sample number eight. Of course, most of my day was spent just locating where everything was and doing the initial prep work for a flow diagram of the processes. Tomorrow, I will be reviewing the steps in the process and examining the test data. I hope that will include the actual data from said life test."

It was Ann's turn. "Sun and I have an appointment with Chief Inspector Ji tomorrow morning at nine thirty. Sun, do you think it would be all right if at least Rick joins us?"

"I don't believe Rick being there will be a problem; however, I would prefer we do not outnumber them during the meeting. Sometimes police in Changzhou can be a little paranoid about situations like that. Sorry, Terri."

"Oh, that's okay with me. I know you three can handle it," Terri replied.

"Sun, can you tell me what happened in your meeting with Chief Inspector Ji today, and also, what do you hope to accomplish tomorrow?" I asked.

Sun began, "Our meeting with Chief Inspector Ji was only a preliminary meeting to establish who I was, who Ann was, and to understand when Ann would be able to see John. I also wanted to prepare Ji that I was going to be pressing him for his evidence against Mr. Alworth and when and if the Changzhou police were going to formally charge John. Chief Inspector Ji agreed Ann could see John today, and I believe he will be in a position to release his evidence.

"Today, then, the first thing on the agenda is to give Ann enough time with her husband. I believe she wants her time with John to initially be private. Right, Ann?" Sun asked.

"I certainly hope so. I haven't seen him in almost a month and have not heard his side of what happened. I know it is also important you all also hear what he has to say, but I need my time first."

"We fully understand and will not intrude on your meeting, but please remember, anything he says to you could be discussed with Chief Inspector Ji at a later date. There are no protections in China regarding a wife and her husband's conversations."

"Thanks for that advice, Sun. Sure glad you told me before not after me seeing John!"

"I think the second most important issue is to find out what the police know about the murder and John's potential involvement," Sun said. "Rick, I certainly want you there for those discussions. We need to be prepared to move quickly if they are planning to formally charge him. Of course, I will push for a speedy hearing and an establishment of bail so we can get him out of there as soon as possible. Mr. Brightson assured me WMT would post whatever moneys were required."

"I am positive Gregory will personally deliver the check if necessary," Ann replied.

"Yes, my being there when the evidence to formally charge John is presented would be helpful first hand for me. I had the opportunity to talk to John in Shanghai, so for right now, I have nothing more to ask him. I know I will at some point after we have developed enough data. Especially if I can locate the additional extended eighth life test."

Sun, Ann, and I agreed to meet in front of Sun's offices at 9:00 on Thursday morning for the drive to the Central Changzhou Police Station. I agreed with Ann for us to meet in the lobby at 8:45, which gave us enough time to walk the short distance.

We left the restaurant at 10:00 that evening with each of us praising our dinners and the wines Terri had selected. Sun, however, made us promise our next dinner together would be Chinese, and he would select the establishment. We agreed not to invite Gregory so we did not have to hear him whine all evening. A very professional and gracious whine, of course!

We three walked Sun to the hotel entrance and his waiting car. We said our goodbyes as his car pulled into the late evening traffic. Terri wanted a nightcap, and Ann agreed. The lobby bar was mostly empty, with only a few businesspeople and an older American couple, so we decided this was the place. We took a seat next to the windows that looked out on the front arrivals area of the hotel and West Yanling Road.

"Rick, I'm scared," Ann said. We had just received our drinks, and the young waitress had left the area. "What if we can't prove John didn't do this; what if the witnesses are right, and they really did see John shoot Mr. Zhu? My God, what would we do?"

"Well, first, Ann, you and I both know John Alworth could never do such a thing, and second, I won't believe anything about a witness until I have my opportunity to talk to the witness directly. According to John, Zhu was alone when he left the QC lab. I believe the assembly line in question had also shut down prior to John's leaving. If that is the case, how could anyone working on the assembly line have seen what happened?"

"I know he couldn't. I guess I have seen too many television reality murder shows and have seen how things can get so twisted and ugly, and he is my husband! God, I miss him."

"Ann, Rick, Sun, and I will not let anything like that happen. You know that," Terri said. "We will find the evidence to prove his innocence and hopefully point the police to the real killer."

"I know you will. I just have to keep believing that, but Jesus, it's my fucking husband we are talking about here!"

With that, we fell into a forced silence and finished our drinks. I wanted to feel sorry for Ann, but something was nagging at me. I didn't know what it was, but I had never felt that way about her before.

Ann's room was on the ninth floor; we said goodnight as she exited the elevator, and I reconfirmed our 8:45 meeting in the morning with Sun.

Our suite was a welcome sight as we slid the entrance card and opened the door. It had been a long day, but a frustrating one as well. I knew Gregory was going to be at the plant sometime tomorrow afternoon expecting some evidence from me regarding the additional life test so he could make a decision on starting production. I really didn't see how that would be possible based on what we knew today.

"Honey, can I ask you a question?" I said.

"If it is about sex tonight, forget it! I am too full and too tired to participate!"

"Well, damn!" I replied. "Foiled again. No, it's not about sex. It's about Ann."

"What about her?"

"I really don't know where to begin. She seems concerned about John and the situation. She says the right words, but she seems almost, ugh, too controlled."

"I'm not sure what you mean by controlled, but you know how she is," Terri said. "She always wants to be the cool, collected one, almost aloof."

"I guess so. Maybe it's because we haven't been around her as much as we normally are, and I have forgotten that. Oh well, no big deal and probably not worth bringing up."

It did not take long for both of us to fall asleep listening to *Saturday's Rhapsody*, our favorite late night CD on Terri's iPod radio.

Chapter 15

Wednesday morning, May 22, began like most days for us. Morning stretch, run, workout, and steam plus breakfast. Little did we know that today would begin a string of events that would bring this case to a close in a way we never anticipated.

"I think I will do a little shopping while you are with Ann and Sun," Terri said. "Anything I can find for you?"

"How about some more energy for you for later this evening so I won't be foiled again!"

"Very funny, smart ass! I'll see what I can do about that, but don't hold your breath! What time do think you will be back?"

"Not sure. I guess around six or so. Why?"

"Well, I suggest you come alone unless you do want to be foiled again."

"Now that you have explained it that way, I will certainly be alone!" I said. Terri smiled and pushed me out the door in time for me to meet Ann in the lobby.

"Call me during the day so I know what is happening," she said. I agreed as I entered the elevator.

Sun and his driver were waiting for us when we arrived at Changzhou, Huaide, and Zhanglinfang. "I don't know why I didn't think last night to pick you up at the hotel rather than have you walk here," he said.

"The walk was no problem. It is a nice morning, and I needed something to get me going," Ann said.

Our trip to the central police station took less than twenty minutes. Sun's driver was unable to park inside the courtyard of the police department grounds, so he stopped outside the main entrance in front of the guard

178

house. We were greeted by a brown and white spotted dog that apparently made his home at the police station because every police officer who left the building stopped to pet it. The dog was completely comfortable with the constant movement into and out of the building.

Changzhou Central Police Station is central to all areas of the city. The buildings reminded me of a corporate headquarters built in the 1970s. The main building was four stories tall. Each story was covered with floor-to-ceiling windows. Portions of most of the windows were open, an indication there was no air conditioning in the building.

A small worn sidewalk swept past the station entrance. A twelve-foot-high cement block wall was painted gun metal blue across the bottom, rising to about three feet. The balance of the wall was painted white. The wall encircled the full property and was guarded at each corner and the middle entry gate. There was a sign hanging across the entry way painted blue trimmed in white saying "police" in English and Chinese. Warning signs and official government documents were positioned on both sides of the entryway. There were four wide marble steps leading to office-type glass entry doors, and the main check-in and waiting area. Marble is big in China!

A station sergeant sitting at an imposing high desk greeted us and asked if we had an appointment with someone located in this particular police station. Sun mentioned later most police departments deal with much more than crime. Many times, they are forced to deal with runaways, tourists, or Chinese individuals not familiar with the city looking for loved ones, directions, or places to stay. Sergeants became tracers of lost persons, city street guides, and recommenders of hotels and boarding houses.

Sun asked for Ann's and my passports as a way to introduce us to the station sergeant. A quick check of the documents and the day's appointment calendar led to a telephone call from the station sergeant to Chief Inspector Ji to inform him of our arrival.

Chief Inspector Ji greeted both Sun and Ann before looking in my direction. Sun returned Ji's greeting, introducing me while describing what part I was playing in the investigation. He also told Ji I was sanctioned by the firm of Changzhou, Huaide, and Zhanglinfang.

Chief Inspector Ji acknowledged both me and my involvement with a firm handshake.

Chief Inspector Ji looked to be about forty-five. He was small in stature with his black hair cut short and wore small hexagon-shaped eyeglasses. I would learn later he was a twenty-five-year veteran of the force and had been promoted to chief inspector three years ago. He was highly rated and respected by not only his peers, but his superiors as well. He was a heavy smoker, which created a problem for Terri and me when we travelled with him.

Chief Inspector Ji was a widower with one son who was finishing university in Changzhou. He was a favorite of the local media since he personally solved one of the most high-profile murder cases and was the star witness during the trial.

We climbed the stairs to the third floor and Ji's office. There was no elevator in the building nor air conditioning in any of the offices. There were two large conference rooms that did have standing room air conditioners for meetings with the press and other dignitaries.

Fortunately, we were considered dignitary enough to merit one of those conference rooms. I, personally, was glad. Spending the morning in a non-air conditioned room or office would not have been productive after an hour.

The conference room was large enough to seat at least twelve people around the square dark green laminated conference table. The top was trimmed with aluminum you would see in many older US factory conference rooms. The chairs were hardwood oak and not very comfortable. There were at least a dozen more of the same type of chairs along all four walls. Windows stretched across the left wall with a view overlooking the main entrance of the police station.

The three of us took seats in chairs at the table with the windows to our backs. Chief Inspector Ji was seated in the middle of the table on the opposite side and looked rather small in comparison to the table, chairs, and room. I saw why Sun was concerned about Terri joining us. One more person would have been overwhelming.

Ji retrieved two large notebooks that were sitting directly in front of him. He handed one to Sun and apologized to Ann and me for not having copies for us, explaining he did not have time to make more than one copy. We assured him not having a personal copy was not a problem, and Sun said he would make sure we had what we needed from the books.

"These books contain all the evidence we have accumulated on this case. I plan to review it with you today and to answer any question you may have. Mrs. Alworth, you have a time scheduled at ten this morning to meet with your husband. It is about that now. One of my assistants should be here momentarily to escort you to the interview room. She will explain our rules to you for these types of meetings and will be with you the entire time. Please understand you will also be required to proceed through security as a safety precaution. This is done with everyone who is allowed in the interview room area."

"I understand. A security check is a small thing compared to the length of time since I have seen my husband," Anne said. She had just finished her comment when the door opened, and a young girl dressed like a detective entered the room and spoke to Chief Inspector Ji.

"Mrs. Alworth, this is Miss Ting; she will be your escort," Ji said.

"Thank you," Anne replied.

Miss Ting and Anne left the room, and Chief Inspector Ji began, "May we please begin our discussion regarding the data in the book?"

"Yes, we are ready," Sun responded. "Do you prefer questions as we progress or at the end of the discussion?"

"Please ask any questions you may have as soon as you see the need."

"Thank you."

Chief Inspector Ji opened the case book and began the discussion. "Our first indication there had been a murder at UCPC came with a telephone call to the police station by the plant manager. He told the station sergeant his name and the name of the business. He said that someone was dead in the quality control lab, apparently shot."

Sun asked, "When did the call come in to the station?"

"Our telephone records indicate the call came in at six thirty-three p.m. The sergeant on duty requested the person making the call identify himself. The caller identified himself as the plant manager at UCPC and gave his name as Peng Wu. We have since verified the person making the call was indeed Peng Wu. He told the sergeant a Mr. Zhu Zhong Huang, the vice president of quality and engineering, was lying dead on the floor of the QC lab. Mr. Wu said he had checked Zhu's pulse, and he had none. Mr. Wu said a night shift cleaning lady had found the body as she entered the lab to do her daily cleaning."

"Do you know if anyone could have disturbed the crime scene before the sergeant received the call?" Sun asked.

"I cannot accurately say that the crime scene was properly secured, because my team did not arrive until after nine that evening," Ji replied.

"What was the condition of the crime scene when your team arrived?" Sun asked.

"I was with the team," Ji said. "The lab appeared not have been disturbed after the murder, except Mr. Zhu's arm had been moved where the plant manager took Zhu's pulse. I knew that because the blood that pooled around Zhu's right arm was smeared across the floor."

"I want to make sure I understand what you just said," I asked. "You said the blood was smeared because of the right arm being moved across the floor, supposedly for the plant manger to check Zhu's pulse? Am I correct?"

"Yes, that is correct," Ji replied.

"Which means the plant manager did not lift Zhu's hand straight up, but dragged it toward him before lifting," I said. "Any idea why he did that?"

"I asked him that question, and he told me he had to stretch his arm and hand to reach across a large pool of blood, and he did not want to disturb it; however, the reach was too far, and he was unable to lift the hand straight up, therefore dragging Zhu's arm into the blood on the floor."

"I suppose that event screwed up any evidence of suicide because the position of the gun had been moved?" Sun asked.

"No doubt the gun was not in the original position; however, Mr. Wu told me that, in his opinion, the gun was lying as if it had fallen out of Zhu's hand."

"Do we believe him?" I asked. "The position of any weapon after anyone commits suicide can vary from case to case and the size of the weapon, plus several other factors."

Chief Inspector Ji responded, "Like you, I also have seen positions of death objects during my career, so nothing surprises me anymore, but I would agree with you the movement of the gun in this case does make it difficult to prove Zhu may have committed suicide."

Ji continued, "Of course, we have no reason to believe he would have done such a thing."

"I understand," I said.

"We also have the statement from the cleaning lady, a Mrs. Xiaojia Dong. She claimed the body of Mr. Zhu was lying in the middle of the floor next to the coordinate measuring machine. That there was blood on the floor around him and a gun on the floor next to him. She said she screamed and ran from the lab to get her supervisor, Mr. Tian. Mr. Tian was in another building, so she found the plant manager. The plant manager accompanied the cleaning lady to the lab, where he found Mr. Zhu, checked his pulse, and called us. The sergeant on duty requested Mr. Wu and everyone else please leave the QC lab and prevent anyone else from entering. Wu agreed to do so and, as I said, it appeared that had been done upon our arrival."

"Have you determined the time of the murder, Inspector?" Sun asked.

"Our forensics team identified the time of death between five and five thirty."

Chief Inspector Ji continued to describe the scene of the murder. "It appeared Zhu had been shot in the chest at close range and had fallen backward, hitting his head on a very large and substantial machine. The lab determined there were powder burns on the front of Mr. Zhou's light blue lab coat, so we believe the shooter fired at close range.

"The shot caused the fall backward, and hitting the machine had no initial effect. There were no signs of a struggle. I would suspect Mr. Zhu knew the killer and was not alarmed prior to the shot being fired. The machine is positioned on the left side of the lab in the middle. I believe Zhu recognized his killer and was facing him when he was shot. The shot entered his body with such force he was propelled backward toward the machine and hit his head when he fell.

"The gun, a Smith and Wesson Chiefs Special, Model CS45, has been identified by several people as the gun Mr. Alworth carried in his briefcase."

"Have you established that the gun on the floor identified as Mr. Alworth's was indeed used to commit the murder?"

Chief Inspector Ji said, "We were able to remove the fatal bullet from the body and match it to other bullets left in the gun. The bullets were identical and therefore a perfect match."

"Inspector Ji, I assume Mr. Alworth's fingerprints were on CS45?" I asked.

"Yes, Mr. Alworth's fingerprints were all over the gun, including the trigger area."

"And no others were found?"

"That is correct."

"Were there any other fingerprints found in the QC lab that did not belong there?" Sun asked.

"No, nothing out of the ordinary. Just prints of individuals who frequent the QC lab as part of their normal working day.

"Which says Mr. Alworth could have returned to the plant, taken his gun from his briefcase, shot Zhu, and fled the scene with enough time to take the six twenty-one train to Shanghai. There is no question Mr. Zhu would not have been concerned to see Mr. Alworth back in the plant and could have been facing him," Inspector Ji said. "We believe something or someone startled Mr. Alworth, and he dropped the gun and ran."

"Have you any witnesses that saw Mr. Alworth in the plant around the time of the murder or may have seen him running in the plant around five thirty?" Sun asked.

"No one has come forward as of now, but Mr. Da is going to offer a reward to anyone who comes forward with any evidence that will assist in convicting the killer. Once that becomes known, someone will."

"How long does the plant's main gate stay open, Inspector Ji?" I asked.

"It closes down at the end of the second shift, which ends at midnight."

"And how does someone get into or out of the compound after midnight?"

"There is a telephone that alerts a security guard making rounds within the compound that someone is at the main gate," Inspector Ji answered. "That guard opens the gate and records the time of day and obtains the signature of the person or persons entering or leaving."

"I suppose there is a daily log of such entries?" Sun asked.

"Yes, there is a log, and we have copies of all of them on the dates around the murder."

"Is there a similar log kept of anyone entering and leaving the compound during normal business hours?" I asked.

"Unfortunately, no," Inspector Ji answered.

"That is unfortunate," Sun said. "Any idea why?"

"I questioned the plant manager and Mr. Da, the owner, and neither one could fully explain why no log was kept during normal business hours. My guess is it had just never been done."

"So, in effect, someone could have come into the compound during normal business hours, removed Mr. Alworth's gun from his briefcase, waited until everyone had left the plant area, killed Mr. Zhu, and left the plant prior the end of the second shift," I said.

"And I guess one of our famous Chinese pandas might also have wandered into the plant, found the gun, and while playing with it, shot Mr. Zhu, but I don't think so!" Inspector Ji said. "I believe it is clear who is responsible for Mr. Zhu's murder. We just need to tie up a few loose ends prior to formally charging Mr. Alworth."

"Not so fast, Inspector, I believe there is still a small problem with a motive and any evidence of the additional extended life test proving there is something fatally wrong with the product. Until these two things have been proven, I believe John is innocent, and I intend to prove it!"

"Okay, okay, I believe both of you have made your positions very clear," Sun said, trying to become the mediator. "Inspector, unless there is anything else you would like to tell us, I believe we have enough information from our discussion today and your fine report to allow us to do some of our homework in this case."

"No, so far, now, that is all," said Chief Inspector Ji. The tone of his voice was enough of an indicator that it was best to end the meeting.

We said our goodbyes to Chief Inspector Ji with a thank you for sharing the information so quickly and openly, and agreed to meet later as events dictated.

Inspector Ji asked us to wait while he telephoned the main desk to see if Mrs. Alworth had finished her visit with John. She had and was waiting for us in the lobby area. Sun requested time for him and me to meet with John tomorrow, and Inspector Ji agreed. We arranged for the interview at 10:00 the next morning. Everything completed, we told Inspector Ji we would show ourselves to the lobby.

I saw Ann was outside talking on her cell phone, so we thanked the desk sergeant as we exited the building. Ann hung up as soon as she saw us coming toward her. "How did your visit go?" Sun asked.

"God, it was good to see him, but he doesn't look good. The bastards aren't giving him enough to eat, and this not knowing what is going to happen is driving him nuts!" she replied. "We have to do something. We just can't let him sit in there and rot! Is there any way we can get him out of there? I know Gregory has more than enough money. Who do we have to bribe? No offence, but you do know how money talks in China."

"No offence taken, and yes, money in some cases does talk in China," Sun replied. "However, this case is too high profile for that type of thing to happen. There is nothing we can do until formal charges have been placed, and from our meeting this morning, Chief Inspector Ji is ready to do just that. I believe all he was waiting for was to have one final meeting with us."

"Jesus Christ, I hope so. At least then we will be able to get John out of there, and he can help us prove he is innocent," Ann said.

"It's eleven thirty now; why don't I call Terri and have her meet us for lunch so we can discuss where we go from here?" I said.

Both Sun and Ann agreed. Terri had just returned to our suite when she answered my call. We agreed to meet here in the lobby restaurant of the Grand at noon.

Terri was waiting for us in the lobby of the hotel when we arrived at just past noon. The restaurant was not very busy, so we selected a table facing the street and ordered almost immediately after being seated.

"How did the shopping go?" I asked Terri.

"Well, let's put it this way: our bank account is somewhat lighter than when I started."

"Go, girl. Spoken like a real shopper," Ann said. Which really wasn't normally true—Terri hated shopping and did most online. She did have one weakness. That was Chinese anything, so I had expected our trip back home, when we did return home, would be a challenge logistically.

Sun began, "As I said earlier, I believe the chief inspector believes he has enough evidence to proceed with formally charging Mr. Alworth with the murder of Mr. Shu and will do so within the next couple of days. As soon as that happens, I will request bail. I assume from comments made by Mr. Brightson that he will personally guarantee John will not leave the country and pay whatever is established as the bond cost? Am I correct?"

Ann replied quickly, "Yes, Gregory, Mr. Brightson, will most certainly do both."

"Okay, then we wait for that to happen and proceed. Please make sure Mr. Brightson stays in country and available."

"I have a meeting with him later today by telephone to discuss where we are with the case and to help him decide whether or not to release the product for production," I said. "I don't believe I am ready to have that discussion today, but hope to have enough data after a full day in the plant tomorrow. I will convey your instructions to him during that call."

"When is that call, and what are our plans for the rest of the day and tomorrow?" Terri asked.

"The call is scheduled for four p.m. I was going to call him on my cell from the plant unless any of you would like to be on the call. This afternoon, I plan on returning to UCPC and working with Billy on the test data and continue my search for the elusive additional extended life test. Finding and verifying that test may be the key to proving John is innocent," I said.

"I believe Terri and I should join you at the plant and complete our task of interviewing everyone that has any possible connection to the murder, but we really don't need to be on the call. Okay, Terri," Sun said.

"Perfectly fine with me," Terri replied.

"I don't believe you need me for the next few days, so I am going back to Shanghai to try to get my business back in order. I can be here on the next train when you call." Sun provided a car for Ann to return to the train station. I called Mr. Da, who invited Ms. Xu to join us to translate. I informed him we were returning to his plant that afternoon as scheduled and arranged for a quick meeting upon our arrival.

Terri, Sun, and I met with Mr. Da and Ms. Xu. Mr. Da listened politely as we updated him, through Ms. Xu, on the events that had occurred since yesterday. He thanked us for the information and asked that we move as quickly as possible to release the new product for production. He reminded us he had a substantial amount of investment in the startup and needed to recoup his money through production orders.

We said we understood his dilemma and would operate as quickly as possible. We also indicated that whatever actions the chief inspector took would have no bearing on our investigation and report to Gregory Brightson. I assured Mr. Da I would advise Mr. Brightson as soon as possible of my findings working with Billy on the life test data and the search for the additional extended life test data. Mr. Brightson would make his decision using my information.

I also told Mr. Da I would be talking to Mr. Brightson late this afternoon and recommend Mr. Brightson give me one more day of investigation prior to making a decision on production.

Mr. Da said he understood and would await Mr. Brightson's decision. We ended our meeting and left the conference room for the plant.

Ms. Xu accompanied Sun and Terri to complete the interviewing of the assembly workers and the cleaning staff while I went to find Billy.

I found Billy in the quality control lab doing first piece measurements of a bracket used on one of UCPC's parts for another customer.

"Hey, Billy, how are you today?" I said as I entered the lab.

Billy took one more reading and recorded the data prior to answering me. "Afternoon, Mr. Watson."

"Billy, do you have time to help me work on the life test data and to search for the missing additional extended life test on sample eight this afternoon?"

"Sure, Mr. Watson, let me finish measuring this part. I only need five minutes."

"No problem. I'll bring up the flow chart and the raw data files in the meantime."

It took me about the same five minutes that Billy requested to assemble everything we would need to continue our investigation.

I wanted to begin by using the new product introduction flow chart to review the first piece of each of the new product components and the first proof of the interface of the WMT software and the UCPC hardware, but I knew Gregory needed to make a quick business decision, so I decided to concentrate on the actual life test data from the initial seven tests. I hoped I would find what I was looking for and could at least give Gregory a feeling of comfort regarding the validity of the planned life test. He could make the decision from that data I supplied.

I returned to the four-drawer filing cabinet in the QC lab where the raw documents were filed and the set of relevant files were on Mr. Zhu's computer. Billy again indicated that to his knowledge, these files contained all of the data regarding WMT's new product.

We had reviewed a substantial portion of the files yesterday, so we were ahead of the game in my preparation for getting the data in order to help Rick make his decision.

WMT had flown seven water samples from key cities in the United States to ensure the validity of the tests. The water processed through filtration systems in China contained different elements than in the United States and could not be used for life tests. Even though in the United States, there are federal and state-mandated specifications on the amount and type of foreign bodies allowed in our drinking water, there were small variances from city to city and state to state.

UCPC had stored the water samples in conditioned rooms to protect the original purity of the sample water. Each sample was one hundred gallons with each container for that particular sample identified. WMT and UCPC hoped to pinpoint any changes to the quality of the water as it was used in the system and through the waste burn process.

I asked Billy to take me to the manufacturing floor so I could see the equipment in person. There was enough water from one sample to run the product as a demonstration, but I did not plan to do so at this time. I was more concerned to see what the equipment was like in person. "So, this is the Master Clear Water Filtration System, is it?" I said to Billy. I was surprised by the size and apparent simplicity of the product. Dennis Dureno's description and the media pictures Terri and I saw during his briefing to us in Shanghai did not do the product justice, and it was easy to see this product would be a bombshell on the industry when it was released.

I could also better understand the need to keep the production and pre-introduction quiet and Gregory's need to make a decision in the next few days. There was a lot to review in a short time, so Billy and I headed back to the lab to continue reviewing the life test data.

Billy and I continued to review the data, but it was obvious there would not be enough information gathered and analyzed prior to my call to Gregory at 4:00 p.m., so I began to formulate in my mind what approach I was going to take to hold Gregory off for one more day. I had a high confidence an answer could be provided within the next twenty-four hours that would allow Gregory to make a go/no go decision on starting production, also in time for the WMT board meeting in Charlotte on May 26. Sun and I were talking to John on Friday, May 23, at 10:00a.m. I would be able to include anything from that discussion in my conversation with Gregory as well. I would be returning to the plant in the afternoon to continue working on the life tests with Billy.

We stopped reviewing data about 3:45 so there would be enough time to return to a conference room in the main office building and for me to call Gregory on time at 4:00 p.m. Billy said he would clean up and organize for the next afternoon. I asked him to wait because we had until 6:00 to work. Billy agreed.

"Hello, Gregory, I trust you have had a productive day," I said as Gregory answered the phone on his end. "You certainly got an early start."

"In fact, it has been a good day," he replied. "I have been going over some numbers with Dennis Dureno on potential business in Asia, particularly China, once the product has been launched in the United States and we are up to production quantities. My God, this could be huge, and we haven't even begun to talk about Europe and South America yet."

"Is Dennis in the United States? I understood he was still in China visiting with local government officials on some technical issues surrounding the product."

"Yes, you are correct; he plans to return to Charlotte next week. What have you learned today?"

"Not as much as I had hoped," I said. "I worked with Billy, Mr. Zhu's assistant, today. I have searched the entire QC lab for the extended eighth life test and have found nothing. I have reviewed all the life test data from the first seven life tests and have found nothing out of the ordinary. Everyone is in spec. Billy and I have begun to assemble the data and format to review the eighth one up to the end of the original life test. I'm sorry I don't have better news."

"I guess I am not surprised. I was wishfully hoping you would be immediately successful, and we could get this portion of the problem behind us and concentrate on John," Gregory said. "When do you believe you might have the data we need to prove our case one way or the other?"

"I hope prior to the board of directors meeting on Monday," I said. "If not, you may have to make the decision without a full compiling of data."

"That I am willing to do if necessary. The company has too much at stake to make a hasty decision if a few days will generate the right data," Gregory said. "Let's talk again tomorrow and see where you are."

"Okay, talk to you then," I said. "I will communicate our discussion to Mr. Da before I leave the plant."

"Thank you, Rick, I appreciate that."

Terri, Sun, Ms. Xu, and Mr. Da were waiting in the main conference room when I arrived. I apologized for being late and explained I had been on the telephone with Mr. Brightson.

"I informed Mr. Brightson I was unable to locate the extended eighth life test, that the other seven life tests all passed life testing, and Billy and I were almost ready to begin a ninth life test," I said to Mr. Da.

Ms. Xu translated, "*Li Cha Xian Sheng Liao Jie Ni He Ge La Ge Li Xian Sheng Tan guo Zan ting Sheng Chan Zhi Dao Shou Ming Shi Yan Jie Shu De Jue Ding.*"

Mr. Da nodded his understanding and agreement.

Sun said both in English and Chinese, "I believe tomorrow will be an important day for UCPC and WMT. Let's hope Mr. Brightson makes the right business decision."

We all agreed, and Sun closed the meeting, promising I or Mr. Brightson would inform Mr. Da of the decision on starting production as soon as possible.

We said our goodbyes and walked from the conference room to the main courtyard where Sun's driver was waiting to drive us back to Changzhou. Once we had waved goodbye to Mr. Da and Ms. Xu, and had entered the traffic pattern, Sun asked, "Rick, what do you think Gregory will actually do regarding starting production?"

"I am not sure. All I know now is Billy and I have not found any evidence of an additional life test being run."

Terri planned to spend the morning going over the interviews of each employee, looking for any new clue we may have missed.

The desk sergeant must have recognized us as we entered, because Chief Inspector Ji met us in the lobby before we completed our security forms and received our stick-on badges.

"Good morning, gentlemen, I trust your trip here was uneventful?" Chief Inspector Ji said.

Sun answered in Chinese that our trip was uneventful, and then switched to English. "Thank you for allowing us a portion of your valuable time this morning for us to speak to my client."

"It is my pleasure, Mr. Sun; however, I must inform you as Mr. Alworth's counsel, I will be pressing formal charges today charging your client, Mr. John Alworth, with the murder of Mr. Zhu Zhong Huang," Inspector Ji said to Mr. Sun.

Mr. Sun showed no emotion or surprise at Chief Inspectors Ji's statement. He said, "You are well within your rights to do so, Chief Inspector. Of course, I disagree with your actions and will proceed to request bail immediately for my client."

"And I believe we have our killer right where we want him, here in the Changzhou Central Police Station," Inspector Ji responded. "I will recommend that no bail be set in this case because of its high profile and the fact that your client is an American."

"Chief Inspector, you know as well as I do that my client will be going nowhere as long as you have his passport."

"Ah, you may be correct, Mr. Sun, but we both know there are many ways to leave China without, let's just say, proper papers."

"And my firm would never be a party to such a thing!" Sun sounded a little annoyed as he answered Chief Inspector Ji.

"Mr. Sun, you and I are not going to solve the situation standing here. Let's let the judge decide Mr. Alworth's fate. Won't you please follow me."

We followed the chief inspector to the fourth floor where the prisoner interrogation rooms were located as well as individual meeting rooms. It was here prisoners would meet with their loved ones and legal counsel.

Sun and I were processed through security one more time and escorted to waiting room 2 where John was sitting. The tables were aluminum with black linoleum tops and aluminum straight-back chairs with matching black seat covers. The mid-morning sun was directly behind John, giving him that angel glow the sun causes.

"Long time no see, Rick," John said. "Where have you been?"

"Trying to prove you're innocent," I responded.

"Well, you are way behind schedule, so I suggest you work weekends," John said.

"You must be Mr. Sun. You are just as Ann, my wife, has described you. Please keep my friend here in line while he is here. He may be sloughing off so I can't be in the states next month to kick his ass during our charity golf tournament."

Sun was not sure how to respond to the trash talk going on at first, but must have figured it out and just laughed at us. "Mr. Alworth, please do not worry. So far, he has been a model companion."

"Terrific, now let's get to it. What is happening out there?"

Sun took the opportunity to inform John formal charges would be made this afternoon charging him with the murder of Mr. Zhu.

"That's crazy," John said. "I had nothing to do with poor Zhu's death. He had really become somewhat of a friend from all of my trips to UCPC. The police are spending too much time trying to prove me guilty while they are letting the real killer get away."

"I understand your position, Mr. Alworth, but you do have to admit there is a substantial amount of circumstantial and real evidence that has been mounted against you," Sun said.

"Look, I'm paying you to not let this happen, Sun!" John was almost yelling when he finished talking.

"Okay, you two, this is getting us nowhere," I said, looking directly at both Sun and John. "All of us are working day and night, looking for evidence that you are not guilty, but we have a long way to go to dispute several of the facts gathered by the Changzhou police."

I continued, "John, Sun is going to request bail as soon as the charges become formal. In the meantime, we need to ask you some follow-up questions based on our investigation."

"Shoot."

Sun began, "Mr. Alworth, some—"

"Please, call me John."

"All right, John, some of these questions may have been asked before. Please do your best to answer as you did then."

"Okay."

"Was your gun in your briefcase in the quality control lab on the day of the shooting, and did you always carry the gun with you?"

"Yes," John said. "The gun is always in my briefcase and was there in the case that day. I have carried it so long, sometimes I forget that I have it."

"Why do you carry a gun, Mr. Alworth?"

"When I first began to travel worldwide for my first company, many of the places we were producing or had customers were sometimes not the safest places in the world. I bought the gun for protection."

Sun asked, "Have you ever used it for that cause?"

"No, never have."

"Please tell me about the argument you had with Mr. Zhu."

"When I arrived at the plant, I met with Mr. Da and Ms. Xu. I told him I was there to review the life test data for the seven sample runs, and if all went well, I planned to release the product for production."

"What was Mr. Da's response?" Sun asked.

"Of course Mr. Da was happy. We had given him an opening production schedule, and he had purchased the inventory for that schedule plus other long lead items."

"Do you know the value of those items?" Sun asked.

"No, I don't, but someone in purchasing in Charlotte could tell you."

John continued, "After my meeting, I went directly to the QC lab to meet Mr. Zhu and review the seven sample run life tests. Mr. Zhu did not seem to be his usual jovial self. We have worked closely together for about three years, so I have or had developed a pretty good relationship with him. He had never married and really didn't have many friends. We had dinner several times while I was at UCPC."

I interrupted John to ask, "Did Mr. Da know how many dollars were tied up in inventory and work in process in his plant?"

John answered, "I'm sure he did. He attended all the progress meetings and signed off on the milestones."

John shifted back to his meeting with Mr. Zhu. "Zhu waited until four of the life tests had been reviewed before he dropped the bomb on me. He told me he had concerns after processing all seven of the life tests and decided to run an eighth life test with higher test hours than our specification called for. Zhu told me there was a catastrophic failure during the last hours of the test."

John said, "It took a few minutes for me to comprehend what he was saying. That just couldn't be possible. Our design engineers have been working and proving this product for three years. It passed every test imaginable in our lab in Charlotte.

"I asked, 'What do you mean you ran an eighth test without my authorization!' You had no right to do so, and how do I know your test was valid? What parameters were used?'Zhu answered before I got my next question out. He said he was afraid to discuss the eighth test in the plant.

"That pissed me off! I guess I was screaming at him. Here we are on the brink of moving to production and a US introduction, and this asshole blew everything sky high and wouldn't talk to me about it!" John's voice began to rise as he was talking, re-experiencing his fury during the initial talk with Mr. Zhu.

"I exploded all over him, and he began to fight back. A lot of words were said that neither one of us should have said, and I am sure you could hear us all over that section of the plant.

"After several minutes, we calmed down and both apologized. I agreed to meet him in my hotel room later that evening."

Sun asked, "Why your hotel room? Why not somewhere in Changzhou?"

John answered, "Zhu said he did not trust anyone here and wanted to get away where he could watch if someone was following him. I didn't know what to think about that—didn't try to, either, just agreed to meet him."

I asked, "Where was Billy when all this was going on?"

John replied, "Not sure where Billy was. I did not see him any during that day."

"Do you believe Billy or anyone else knew about this eighth life test?" I asked.

"Not the way Zhu acted. The man was scared shitless."

Sun asked, "What time did you leave the plant?"

"Since there was nothing more to do at the plant, I left about four thirty that afternoon."

"I thought you had a six twenty-one train back to Shanghai? Why leave so early?"

John answered, "I was tired from the flight over and wanted to get some rest before Zhu arrived at my hotel."

"But you didn't take the six thirty train, did you?" Sun asked.

"No, I arrived at the station about five o'clock and found a seat on the five fifteen train," John said.

"Wait a minute." I asked, "If you were on the five fifteen train, you are in the clear. That's a perfect alibi."

John answered, "It would be if I could find my ticket, but I can't. I have searched everywhere for the son of a bitch, but it's gone."

I said to Sun, "Can't we get the records of tickets purchased to prove John was on the train?"

"It is not unusual for individuals to purchase several tickets when their plans are not firm and switch them around later. Nothing there for us."

"Wait a minute, you have to run your ticket through the machine to leave the station, don't you? Wouldn't John's ticket show up at the Shanghai terminal exit?"

Sun responded, "I have dealt with this many times in other cases and have discovered there are many ways to leave the terminal without sliding your ticket through the exit point. Can't use that in court."

"What about your six twenty-one ticket?" I asked. "What happened to that one?"

"I traded it in on my five fifteen ticket," John said.

I looked at Sun and his return look told me that checking on that was not a good option either.

I asked John, "Do you have any idea where Zhu could have kept the results of the eighth life test?"

"No, not unless he hid it somewhere in the QC lab, but I imagine he took it off the property."

I asked Sun, "Do we know if Chief Inspector Ji's team searched Zhu's apartment?"

"I don't know, but we will ask him prior to our leaving."

"John, you say you left the plant at four thirty, correct?" I asked.

"That's what I said the first time you asked me," John replied.

"Were the guards at the main gate in the guard shack when you left, and did Mr. Da's driver take you back to the hotel?"

"Mr. Da's driver was on the way to pick him up from somewhere, so I took the express bus that runs at the corner. Sometimes, it is the fastest way to beat the traffic," John said. "Busses have their own lanes on almost all the roads back to the hotel. The guards were in the middle of changing shifts and did not pay any attention to me walking by."

"I assume you used change for the bus?" Sun asked.

"That's correct. It's one of the few places you can get rid of it."

Sun said, "John, just a few more questions, okay?"

"No problem, ask a way. Like I have somewhere to go."

"Did anyone else come into the QC lab while you were there?"

"Nobody."

"And Zhu was there when you left the QC lab?"

"Yes, he was."

"And you did not talk to anyone while you were leaving the plant?" Sun asked.

"No, but I wish now I would have. If I had done that, we all would be out drinking a beer right now."

"Rick, any more questions?" Sun asked.

"Just a couple more," I said. "Johnny, did you see anything in the numbers when you reviewed the first four life tests?"

"No, I spent the time working with Billy for the first three and by myself on the last one. Everything was according to spec."

"Where did Billy go for the fourth one?" I asked.

"I'm not sure," John replied. "He said he had a personal problem and needed to leave the plant. He gave me his cell number to call him on if I needed him before he returned."

"Do you think Billy could have the document on the eighth life test?" I asked John.

"I don't see how he could. Zhu kept the large projects to himself and kept things very close to the vest and probably didn't feel the need to discuss it with him."

I turned to Sun and told him I had asked all the questions I had written down and then some. Sun agreed he had also asked the questions he wanted as well. Sun asked John if there were any questions he had thought of that we could answer or ask? John said no, but did say his main interest was getting the formal charges behind him so we could request bail.

We thanked John. Sun promised to call him as soon as the bail hearing had been set. John kept thanking us and assuring us he could handle the confinement until he was bailed out. I was not sure when we would be seeing John next, but hoped it would be within a few days.

Sun and I thanked the guard, passed again through security, left our badges with the guard at the entrance to the fourth floor, and took the stairs to the main lobby. Chief Inspector Ji was waiting for us when we arrived there.

"Mr. Sun, I have just formally charged your client with the murder of Mr. Zhou and am on my way to his cell to inform him," the chief inspector said. "I assume you will be asking for a hearing regarding bail."

"Yes, I will be doing that immediately upon returning to my office."

"Then I guess I will see you in court," Chief Inspector Ji said, laughing as he walked away.

"Thank you for your help, and yes, we will both see you in court."

Sun shouted at Inspector Ji as he walked away, "Oh, by the way, have you or your team searched Mr. Zhu's apartment?"

"No, why do you ask?"

"Oh, nothing. Just curious. Do you plan to?"

"I have no plans to do that," the chief inspector said.

Sun thanked Chief Inspector Ji one more time, and we left the police station. Mr. Sun's driver was waiting for us as we exited the police station.

"Rick, why don't you call Terri and have her meet us in my office. I will have lunch brought in. Terri and I can work the bail issue while you are at the plant. We should be able to have a nice lunch and get you to the plant by one," Sun said.

"Sounds good to me. I will call her right now."

I dialed her cell, and she answered on the first ring. "Hi, honey. I saw it was you on the caller ID."

"Hey, babe, how about meeting us at Sun's office in about fifteen minutes? He is buying lunch, and he wants you to work with him this afternoon on getting John's bail issued. You will need to call Gregory to get the money."

"Okay, see you there in a few."

To his word, Sun ordered an exquisite lunch that we shared in his office around the conference table.

"Sun, how do you do it?" Terri asked. "This lunch is fabulous! A working lunch in the United States is normally an overblown catchy-named sandwich brought in from a local caterer."

"I have a friend who runs a business that specializes in these types of meals. You give him a lead on what your desires are, and he matches it with one of the local restaurants," Sun said. "His team works with the restaurant to ensure the lunch is high quality and produced at just the right time for delivery."

"Honey, we need to bottle that idea and take it back to Charleston," Terri said to me as she was taking her last bite.

The three of us spent the next half hour coordinating the balance of the day before Sun's driver appeared and suggested he and I be on our way to UCPC. I promised Sun I would call him and Terri after my call with Gregory. They also agreed to call me if they were successful in obtaining bail for John prior to my leaving UCPC.

I arrived at the plant right on time and met Billy in the QC lab just after 1:00 p.m. to begin reviewing the next set of life tests.

Our reviews produced no different data than we had seen yesterday. Each life test was performed to the prescribed test plan, and each one passed each milestone within specifications.

"Billy," I asked, "were you involved in each of these life tests originally?"

"Not all of them. My wife was sick for a few days during the initial runs, so I was in Shanghai with her."

"Sorry to hear that. Is she all right now?"

"Yes, she is, Mr. Rick. She had a kidney stone attack."

"If your wife is your age, she is awfully young, isn't she?"

"Her doctor said the same thing. She is on a strict diet now to hopefully prevent another occurrence."

"I have never had one, thankfully, but have known friends who have, and they tell me it is the most pain they ever felt."

"I know my wife told me after it was all over she now knew what childbirth must be like."

"Oh, so you and your wife have no children?"

"No, not yet. She is an engineer with Samson Industries in Shanghai and wants to become more established before we start a family."

"I assume then you go back to Shanghai every Friday afternoon?"

"That's right. Normally, I take the seven p.m. train and return on the six a.m. train on Monday morning."

"I guess that is somewhat normal for married couples in China," I said. "In the United States, we tend to move locations for jobs. Here, one spouse stays in the home area and one travels."

"We are both from Shanghai and graduated from the same university, but I found a job here at UCPC and never thought about leaving until what happened to Mr. Zhu."

"Things will get better," I said. "They always do. We will find the killer, correct whatever needs to be corrected on the product, and production will begin."

"I hope so, Mr. Rick," Billy replied.

We worked until 3:15 and completed our review of all life tests. The trend continued. Each one passed the test specifications at each milestone. There was nothing I found to indicate Gregory should not release the product for production.

The only issue that kept creeping into my mind was the absolute lack of any lead on an eight life test. There was no evidence anywhere in the files indicating there was such a test. I was desperately trying to locate any shred of evidence to support John. He said Zhu told him there was one, and I had no reason to believe John was lying. John's mission in coming here was to approve the life tests and okay production start. He had a financial interest in that happening, so I knew he would lean toward approval, but not if there was any question of the product quality. John's integrity was above reproach.

My cell rang precisely at 4:00 p.m. as I was sitting in the same conference room as yesterday.

"Hello."

"Hello, Gregory," I said. "How are you today?"

"Not very fucking good!" he said, almost screaming into the receiver. "The bastards have denied bail for John, and I am pissed. I just finished talking to Mr. Xue of Changzhou, Huaide, and Zhanglinfang, and he told me they have done everything they can to make it happen. He has personally discussed the situation with the commissioner of police and the mayor of Changzhou, but they will not allow John to leave his confinement. They are concerned he will find a way to leave the country, and they will be left holding the fucking bag! They kept reminding me of the high-profile nature of this case.

"I also personally called both of these guys to assure them John would not leave and be visible to them twenty-four hours a day. Hell, I gave them my personal guarantee, but that did not mean anything to them! Bastards!"

"Well, that's just ducky!" I said. "The one thing I had hoped for was to get him out of there so he could help me find this eighth life test that he swears Zhu told him about that led to the big argument and everything that followed. I guess I will have to work remotely with him from now on," I said.

"One piece of information that sucks is enough today for me. You had better have good news!" Gregory said.

"News I have, but you will need to decide the value and goodness," I replied. "We have found nothing to prevent you from releasing production."

Gregory asked, "Have you located the eight life test?"

"Unfortunately, we have found nothing that would lead us to believe there was an eighth life test, but I trust John with my life, and if he says there was one, there was one," I said.

"I feel the same way. John Alworth is the most honest man I know. I need time to think about this. On one hand, you're right; I should release production. On the other hand, if there is an eighth life test and it is as onerous as John believes, it may be I have an obligation to everyone within WMT to be prudent about a product that could be placed in someone's home and have a catastrophic failure. That could destroy everything we have working for at WMT and potentially cause harm to a customer who trusted us."

"I told Terri yesterday our job of finding John innocent is much easier than your decision about the product."

"That's my job," Gregory said. "I started this company and have driven it to where we are today. I didn't do that by being indecisive, but this one is the biggest challenge I have ever faced. Give me a few hours this evening, and I will give you my decision before bedtime."

"That is okay with me, Gregory. Does Ann know about John?"

"Yes, she is here with me and, as you can imagine, not very happy right now."

"I am sure she isn't. Then I guess I will talk to you later tonight," I said.

We hung up, and I immediately called Terri and Sun. Sun answered, put Terri on speaker, and they basically gave me the same information that Gregory had. I told them Billy and I were finished for the day at the plant, and I should be back at the Grand no later than 5:30. Mr. Da was out of the plant, so I skipped my daily briefing with him. I would bring him up to date tomorrow.

Sun suggested he buy dinner so we could discuss what we should do next. We agreed and decided to meet in the lobby of the Grand at 7:00 p.m. Terri said she was leaving Sun's office and would meet me in our suite.

Terri and I were sitting in the lobby bar nursing a couple of Coronas when Sun arrived. He saw us and made his way across the lobby. "You two look comfortable," he said.

"What does the commercial say? 'After a long day, make it a Corona.' Well, it's been a long day!" Terri said.

"Please, join us," I said. "I'll buy."

"I believe I will, as the commercial also says, 'make mine a Corona,'" Sun answered.

I waved down a passing waitress and ordered our three Coronas. Our beers arrived, and we cheered their arrival.

"How about dinner tonight, you two? Any ideas?" Sun asked.

Terri was quick in replying, "How about Zhangshengji?"

Sun looked surprised and, laughing, answered Terri, "I am very surprised you know the Starway, let alone the Zhangshengji. Most Westerners never find it. This means you have been in China long enough to be considered a local! Actually, it is one of my personal favorites, but I seldom take others there because of the outside appearance. Great suggestion."

Beers finished, we made our way to the Starway Hotel. Dinner was as magnificent as our first visit, and Sun included a few special local dishes to cap off the evening. We spent very little time discussing business, but spent time on personal things, getting to know each other better. This conversation meant our relationship with Sun was reaching more than a normal business relationship, which is difficult to accomplish in China. Most relationships sometimes take years to develop.

We were about to leave our table when my cell rang. It was Gregory. I answered and asked his permission to put him on speaker so Sun and Terri could hear him directly. Gregory agreed and began, "Good evening, everyone. I wanted to inform you of my decision on releasing the product to production before I return to Charlotte tomorrow. I have decided not to release the product into production until we are absolutely positive there is no eighth life test. Rick, I am counting on you and Terri, working with Sun, to prove or disprove this eighth life test. Also, Rick, you will have everything you could possibly need at your disposal, and I plan to be in daily contact with you or return to Changzhou if you need me. I am assuming you will stay on to help both John and I?"

"Gregory, of course. Terri and I plan to stay here until we have proven John is innocent. I appreciate your confidence in us and your offer of resources. I know this was a very difficult decision, and we will support it."

Sun said, "Mr. Brightson, Sun here; our firm and I personally will also commit to working with Rick and Terri in any way they need to prove our case. I will personally talk to Mr. Xue tomorrow as well."

"Thank you, all of you for keeping on top of this. WMT is and will always be a leader in everything we do. I believe, with your help, we will solve this quickly, and I will be able to give my board of directors my word it is time to start production."

"Gregory, when will you be back in Charlotte?" I asked.

"My flight lands late Friday afternoon, and I plan on being in my office early Saturday morning if you need me. Of course, you can always get me on my cell twenty-four hours a day, if need be. My meeting with the board is the twenty-sixth."

"Have you communicated your decision to Mr. Da?" I asked.

"Not yet, but I plan on calling him immediately after we hang up," Gregory said. "Rick, please ascertain his reaction to my decision and let me know how my decision has affected him. He is a valuable partner and personal friend. I will tell him WMT will be responsible for all the inventory and startup costs to date, also additional costs if he has to hold skilled workers assigned to our product for some period of time."

"Will do."

We thanked him, pledging to continue working on the case and agreeing to a discussion at 7:00 a.m. Saturday morning his time and 7:00 p.m. Sunday night our time. We disconnected and sat there for a few seconds, not knowing what to say about the situation or to each other.

Finally, Sun said, "I realize that was a very difficult decision, and I respect him for making it."

"Me, too, but it sure puts the burden on us!"

"I have a feeling you and Terri are up to the task," Sun said. "And remember, I am at your disposal whenever you need me."

I thanked him. We completed our dinner and met Sun's driver in front of the Starway for the short ride back to the Grand in the rain. Sun asked me what our plans were for Friday, and I told him I would like to spend the morning thinking about where we were at this point before making further plans. I promised to call him in the afternoon Friday.

It was after eleven when Terri and I arrived back in our suite. We spent a few minutes discussing our exceptional day and the possible consequences for John due to Gregory's decision. We both felt our next step should be to sort through each of the employee interviews looking for any scrap of information on what happened in the QC lab and the events surrounding Zhu's murder.

I also needed to arrange to meet with John on Saturday and to make a short visit to the plant to meet with Mr. Da, assuming he would be there Saturday morning.

The day caught up with us around midnight, and we both fell asleep listening to the rain hitting our windows.

Chapter 16

Saturday, May 24, dawned later than usual for us, probably because of the continuing rain falling on the city. It was after nine before we made our way to the gym for our normal morning workout. We had hoped for a run through the city streets, but by nine in the morning, everything was in full swing, and the traffic was too busy for a street run, plus running in the rain sometimes is no fun.

We opted for the full breakfast in the main restaurant of the hotel and decided to make lunch some simple fruits.

I called Chief Inspector Ji and scheduled an interview with John at11:00 a.m., which meant I had to hustle to shower, dress, and find a cab to Changzhou police headquarters. I called Mr. Da on the way, but he was out of the plant until Monday morning. Ms. Xu arranged for a 9:00 a.m. meeting with him at the plant.

Inspector Ji was hurrying down the sidewalk in the rain under a rather large umbrella when my cab delivered me to the front of police headquarters. "Good morning, Mr. Watson," Chief Inspector Ji said. "I was out dropping off some papers to one of our prosecutors who is just two doors down."

"Good morning as well, Chief Inspector," I responded.

"Please join me," he said, dipping his umbrella. "It is a wet day in Changzhou."

"Thank you, I believe I will."

We successfully maneuvered our way jointly around several puddles on our way to the front of the building and achieved some semblance of dryness.

Chief Inspector Ji waited until I had signed in with the desk sergeant and accompanied me to the fourth floor where the interview rooms were located.

"I understand Mr. Brightson has decided to hold production on the new product for now; am I correct?" He knew he was correct, but wanted a confirmation from me.

"Yes, you are correct, Chief Inspector. Mr. Brightson is very concerned that we have been unable to locate the eighth life test."

"If there was such a test," he replied.

"Chief Inspector, as I have said, I personally know that Mr. Alworth is an honest man, and if he says there was an eighth life test, there was an eighth life test. I do know the water for the test is not at UCPC."

"Well, I hope for your friend's sake, he is honest the water was used, and he is innocent, but I truly doubt it."

Chief Inspector Ji signaled to the guard on the floor and said goodbye as the guard motioned for me to proceed through security.

John was waiting for me in the same room as the last time, with the guard waiting patiently behind him. The guard stayed in the room with us, and I was positive he spoke enough English to give Inspector Ji a report after the interview was over.

"I hope to hell you can tell me why I am still sitting in this hell hole," John said as I entered the room.

"What, no hello? No 'thank you, Mr. Watson, for all you have done'? No 'how is my best friend holding up'?" I asked.

"How you are holding up! It must be tough living in a suite in the Grand, eating out every night in a gourmet restaurant . . ."

"Hold on, I wouldn't say they were all gourmet," I interrupted him.

"Oh, sorry, what did you do, eat lunch one day in the UCPC cafeteria?"

"Well, in fact, we did," I said.

John laughed and said, "But enough of the bullshit. When are you going to get me out of here?"

"It doesn't look like we are going to until we prove you are innocent, which we will."

"I know. Gregory said as much to me when he called last night. What can I do to help?"

"John, are you sure the guards did not see you leave the plant on that day?"

205

"I know they were busy talking to each other and never looked at me," he said.

"Do you know the names of any of them?"

"No."

"I will talk with Mr. Da to see if we can find the guards on duty that day," I said. "Did you talk to anyone on the train, get the name of someone?" I asked. "How about the conductor? Did you talk to him?"

"No, I got on the train and napped all the way to Shanghai. I was in a window seat, so no one bothered me."

"How did you get back to the Shangri-La?"

"I took the underground, as I normally do."

"Too bad you didn't take a cab just this one time!" I said.

"You know, I have certainly learned something for the next time something like this happens to me."

"What's that?"

John answered, "Seriously, I have been here so many times, I am comfortable using all the mass transit, but using it is not traceable."

"You're right. I never thought about it that way, and by the way, there will not be another time!" I replied.

"Is there anybody but Billy who would have known about the eighth life test? Another quality person or engineer maybe?" I asked.

"I have never seen anybody but Billy and Mr. Zhu involved in things like life tests," John answered.

"Are you sure no one was in the QC lab prior to you leaving?"

"Ann was there early in the afternoon before my argument with Zhu. She stopped by to tell me she was going to stay in Changzhou to meet with a client.

"She asked why I was so mad. I told her, and she told me to calm down and not cause a scene with Mr. Zhu. She liked him."

"Where was Zhu when she was there?"

"He was in a meeting in the main building."

"So, he returned, and that was when you two had the violent argument, correct?"

"That's right. I did my best to remain calm, but just couldn't. Jesus Christ, Rick, our business life was at stake based on what he said! You would have been pissed, too."

"How long had Ann been gone before Zhu arrived?"

"About twenty minutes."

I continued, "You and Zhu argued, you made up, and then you left for the train, correct?"

"That's it."

"No one saw you and Zhu calm down and make up?" I asked.

"No, nobody came in before I left."

I said, "I was going to leave here and go to UCPC to meet with Mr. Da to get his reaction to Gregory's decision and to bring him up to speed on our progress, but he is out of the plant all day today. I have arranged to meet him at nine a.m. on Monday. For now, I plan to work with Terri for the rest of the day reviewing the interview data she and Sun gathered. Not sure we will find anything, but I have to look."

I continued, "Then I am going to push Chief Inspector Ji to search Zhu's apartment for the eighth life test data. I know he will be reluctant, but Sun can help me with that one."

"Good luck, buddy. I know I get impatient and bitchy, but you know how much I appreciate what you and Terri are doing for me," John said.

"I only do it because I want to whip your ass in the golf tournament!"

We laughed, and I made my way back through security. I stopped by Chief Inspector Ji's office, hoping he would be there and available. He was and invited me into his office.

"Chief Inspector Ji, there is something I have been wrestling with in my mind regarding John's case. Do you have a minute?"

"Of course, please," he replied, pointing to his conference table.

"I just need to know, what motive do you believe John had for killing Mr. Zhu?"

"I believe it is very simple. A killing caused by your friend's hot temper. I believe there was an argument, but not over the mysterious eighth life test, and Mr. Alworth went into a rage, pulled his pistol from his briefcase, shot it, and killed Mr. Zhu. Afterward, Mr. Alworth panicked and ran from the scene."

"Why do you believe they argued?"

"I wish I knew. My case would be airtight if I did, but we have several witnesses who will testify your Mr. Alworth and Mr. Zhu had frequent disagreements or I should say arguments during previous visits to UCPC. It is apparent there was no love between them."

I replied, "I am not sure I believe that, Chief Inspector. I know John has a temper and can get really testy with folks sometimes, but I just can't believe he would ever get to a rage deep enough to kill someone. I also know Mr. Zhu was a key individual in the development of the product from an engineering and quality perspective. Sometimes, working together in pressure situations like this does create relationship problems. I have been there myself, so I know how these things can happen, but never enough for someone to commit murder."

"I certainly do not have the experience you have, Mr. Watson," Inspector Ji said, "but I do have considerable experience in solving cases using police investigation tactics, and I believe I am right on this one."

"So, you don't believe there was an eighth life test?" I questioned.

"No, I do not. I believe Mr. Alworth made that up as part of his defense. If there was one, we would have found it by now," Ji said.

"That brings me to my next question and a request," I said. "Why did you not search Mr. Zhu's apartment for such a document?"

"There was no reason to suspect we would find anything there regarding this case," Ji answered.

"I would appreciate it if you would do that now. My friend said there was a test, and I believe him. I have found nothing searching within the quality control lab or the area around it in the plant. Maybe Zhu took the test results home."

"I am always short members of my team who are working on other cases, but I will do so for you."

"Many thanks, Chief Inspector Ji," I said.

I left the police station and hailed a cab. There was no need to go to UCPC. They were not working this afternoon. Most businesses in China initially worked six days a week, but over time, the Saturday workday had been reduced to one half day, giving employees more free time.

I was back at the hotel about noontime. Terri was waiting for me when I entered the suite. After a hug and a kiss, I detailed my meetings with John and Chief Inspector Ji. It was then I noticed my dear wife had prepared lunch for us in the suite. She had been to the hotel kitchen and selected an assortment of fruits and vegetables, along with a bottle of Chinese summer wine.

After a refreshing lunch, we began working our way through the mound of data Terri and Sun generated during the interviews at the UCPC. We had discussed our approach during lunch. We were looking for anything that was said during an interview that would cause us to pause and feel the need for a follow-up discussion.

Into the second hour, the mound had shrunk significantly, and we had only selected one interviewee to re-address. The testimony from the welder on the assembly line didn't feel right. Terri remembered he was hesitant answering many questions and seemed nervous.

The pile was complete by 3:00 p.m. It was still raining the same slow drenching rain of the past two days, cancelling most of our ideas for outside plans for the rest of the day. Terri suggested we visit the Changzhou museum to help relax and clear our minds from the case. It seemed like a good idea to me, and it kept us out of the rain, so I quickly agreed.

The museum was a short cab ride into the western section of the city and People's Square. Completed in December 2006, the museum was opened to the public in April of 2007. The building architect must have admired the main theater in People's Square in Shanghai because the outside architecture of the Changzhou museum was a smaller duplicate.

The building was four stories high with the special display area on the second floor. Travelling exhibits and local classic exhibits were displayed on the second floor by the museum's curators. The first and third floors were used to display the standing artifacts, while the top floor was used for offices. The underground floor was used for storage and to check in, check out, and prep exhibits.

Terri and I spent most of the rest of the day browsing each floor open to the public. Terri wanted me to put a jade piece from the Liangzhu period in my pocket, but I thought not! It was still raining when we exited the museum in the early evening. We hailed a cab and arrived back at the Grand around 7:00 p.m.

Saturday night was quiet with dinner in the hotel and an early bedtime, again with reading and listening to the rain as we fell asleep. Sunday was a day of rest for both of us. We needed time to ourselves to decompress from the past seventeen days. Next week would be a critical week for us and John. The Sunday morning buffet was just too beautifully presented to miss, so we splurged by eating too many of the items on the buffet and going completely off our normal diet. We promised ourselves to run further and exercise harder on Monday morning to compensate, and besides, if we had a large brunch, we would skip lunch.

Chapter 17

We awoke Monday morning, May 26, to bright sunshine streaming in our windows, a welcome sight after almost three days of rain. It was refreshing to run in the city after it had been cleansed by the rain. The sidewalks were streaming as we discovered new areas of Changzhou not too far from the hotel. People of all ages were also enjoying the morning with us, doing tai chi in small parks along the way. We also ran through, by mistake, a Chinese exercise dance class. We stopped, bowed, and apologized before we sprinted away laughing at what we had done. The dancers seemed to enjoy the interlude more that we did, sending us off with waves and returning our bows. The rest of the workout and breakfast didn't hold a candle to the morning run!

Mr. Da's driver was waiting for me outside the hotel at 8:30a.m. The ride to UCPC was unusually quick.

UCPC was fully staffed and operating like any other normal weekday of production when I arrived. Mr. Da's driver called ahead to inform Mr. Da of our arrival time, and he and Ms. Xu were waiting for us when we pulled into the courtyard.

Mr. Da bowed as I exited the van and said in English, "Good morning, Mr. Rick. Welcome back UCPC."

"Good morning to you as well, Mr. Da, and good morning, Ms. Xu. Many thanks to both of you for spending a part of your day with me." Ms. Xu translated for Mr. Da, who smiled and said, "Most happy to see you."

Ms. Xu suggested we meet in the normal conference room and led the way there.

When we were seated, I began by thanking Mr. Da for his kindness for meeting me this morning and promised to be brief. Ms. Xu translated to Mr. Da who replied, "Ah, thank you."

I was not sure he understood, but plunged ahead with my information and request for information from employees who worked in his factory.

Mr. Da nodded his head and smiled. Ms. Xu assured me he understood and was willing to spend whatever time I required.

I was about to begin when a young assistant entered the room with hot tea and juice. I opted for the hot tea while Ms. Xu had juice. Mr. Da had his usual special tea blend.

My first mission was to gauge his reaction to the telephone call from Gregory. I asked, "Mr. Da, I understand you have talked to Mr. Gregory about his decision to hold production until we have proved or disproved there was an actual eighth life test?" I asked.

Ms. Xu translated: "*Shi, Wo He Wo De Peng You Ge Lai Ge Li Xian Sheng Tan Guo. Ta Gao Su Le Wo Ta De Jue Ding. Wo Bu Gao Xing, Dan xin Shang Ta You Guang Cai Liao He Gong Ren De Ti Yi.*"

Mr. Da replied, "*Li Cha Xian Sheng Gan Xie Nin Li Ci Shi Bing Qie Bang Zho Wo Men Xun Zhao Dao Zhen Zheng De Xiong Shou, Ru Ni Suo Yan.*"

Ms. Xu translated, "Mr. Da say he talked to his good friend, Mr. Gregory. He say he not happy, but appreciate his offer for material and workers."

"Thank you from me as well for you being so understanding and helping us find the real killer," I said to Mr. Da.

Ms. Xu said, "*Li Cha Xian Sheng Shuo Ta Jiang Wen Ni Ji Ge You Guan Mou Sha Dang Tian De Wen Ti, Hao Ma?*"

Mr. Da replied in English, "Ah, okay for me. I, too, want to find real killer, as you say."

I told Mr. Da about the conversation I had with John Alworth regarding his leaving the plant on the day of the murder and no one in the guard house seeing him. I asked Mr. Da if he could find out for me who was on duty that day about 4:30p.m.

Ms. Xu quickly translated my request, which was followed by several minutes of discussion and two telephone calls. Finally, Ms. Xu said, "We have knowledge of which guards were on duty at four thirty p.m. We will talk to them and let you know the results of our discussions later today."

I thanked Mr. Da and Ms. Xu for acting so quickly. I also told them I would like to re-interview one employee on Tuesday.

Ms. Xu asked which employee I needed to talk to. I told her it was the welder on the assembly line. She discussed this with Mr. Da, and he replied, "Please, Mr. Rick, you may talk to him."

I asked Mr. Da if he had any more questions for me, and he replied he did not. I thanked them for their time as we made our way to the front of the main building and the driver waiting to take me back to the Grand. Billy was in training for the day, so I decided to return tomorrow to continue.

They both bowed, and I waved goodbye as the van left the courtyard into city traffic. I was back in our suite by 11:30.

Terri was not in the room, but left a note saying she was in the pool and to join her. Sounded inviting to me! She did not see me coming, so I did a cannonball right behind her.

"Are you crazy!" she yelled as water engulfed her. She spun around, ready to give some thoughtless stranger a piece of her mind, when she saw me surface.

"I should have known!" she said, laughing. "No self-respecting Chinese person would do such a thing. Just a rude American!" With that, she splashed me and dove on top of me.

The pool was as refreshing as I had hoped. We spent a couple of hours just relaxing both in the pool and hot tub.

"How did the meeting go?" Terri asked.

"Actually, much better than I had expected. Gregory did a smart thing by offering to pay for inventory and the wages of the workers until he makes a final decision."

"How about our request to re-interview of the welder on Monday?"

"No problem."

Neither of us wanted to venture far from the hotel, but Bin Wang called. He was in Changzhou, apologized for such a late call, but said if we had no other plans for the evening, he would be pleased if we would join him for dinner. We had not seen or talked to Bin since our first days in China. Dinner tonight would be an excellent opportunity to catch up. Bin, being Bin, deferred to Terri to select the restaurant. She picked the Zhangshengji. Bin said he had not had the pleasure of dining at Zhangshengji, so was willing to try something new.

The three of us spent the evening enjoying another incredible dinner and discussing all the events surrounding John Alworth and are view of what was happening in Bin's life.

Terri and I agreed it was going to be hard to leave Zhangshengji when the case was over. Maybe we could open a duplicate Zhangshengji in downtown Charleston with the same name and pretty much the same menu. We could call the food Southern China cooking!

Chapter 18

Tuesday, May 27, was another beautiful day; however, summer humidity was on the rise. Humidity in this part of China was normally more oppressive than in Charleston. I truly hoped the case would be solved and we would be back in Charleston before summer Chinese humidity kicked in. We repeated our run outside, this time, however, avoiding the dance class. We ate breakfast in the executive lounge after our stretching and sauna.

I had scheduled a 9:00 a.m. call with Gregory to discuss Mr. Da's reaction to his decision to hold production and to report any new developments on the unapproved life test.

I was reading a John D. McDonald novel while waiting for time to make my call and was deep into the plot when I heard Terri say, "It's almost nine o'clock, honey. Do you still have a phone date with Gregory?"

"Yes, I do."

"Have you prepared what you are going to say to him?"

"Not much new to tell him. I plan on bringing him up to date on my meeting with Mr. Da and his reaction to his decision, that we still have no evidence of the eighth life test, let Gregory know we have completed our analysis of each one of the employee interviews and have selected one person to re-interview, and my request for Chief Inspector Ji to do a search of Zhu's apartment."

Gregory answered on the first ring. "Hello, Gregory, how are things in Charlotte this morning?" I asked.

"Actually, I have been in my office for about three hours," he replied. "I have a significant amount of e-mails, texts, and regular phone calls to read or return. My presentation to the board of directors is almost complete. I was hoping maybe you had some good news for me so I could scrap the whole thing and start over on a more positive presentation."

"Sorry, Gregory, no such luck right at this moment," I said. "I wish I also had better news, but not much has changed." I spent the next several minutes repeating what I had said to Terri just before I made my call.

"I am pleased Mr. Da is okay and accepts my decision. I know that was a bitter pill for him to swallow; that's why I agreed to pay for the current raw material and in-process inventory, plus the salaries of key employees who are attached our Master Clear Water Filtration System. Once this problem is solved and production begins, it will become critical to have everything in place at UCPC, as well as here in Charlotte, so we can ramp up quickly."

Gregory continued, "I agree with you it is a good idea to search Zhu's apartment. We may not find anything there, but we must be sure there is nothing there that would be detrimental to John or something that would injure WMT. If we do find the so-called eighth life test, and it does contain catastrophic information, at least we can attack the problem and correct it. Also, I like re-interviewing the welder on the assembly line. If Terri and Sun were suspicious, then we have nothing to lose by talking with him."

"Thanks for understanding the situation," I said. "I truly believe we will achieve a breakthrough early next week. I can just feel it. We just have to be missing something big, but I can't put my hands on it. I know the solution is out there, waiting for us to finally wake up and recognize it. I promise you that Terri, Sun, and I will not quit until the real killer is unmasked. I appreciate your confidence in Terri and me. Like you, we believe John is innocent." I continued, "All we have to do is track down the real killer and verify there was an eighth life test so WMT may deal with it."

"That's still a big order, but I, too, believe we are closer to solving it. I don't mean to be rude, but if you have nothing else, I think I need to get back to work," Gregory said. "Call if anything earth shattering happens before Monday morning."

We both hung up, and I turned to Terri. "I'm sure glad it's not me going before the WMT board on Monday," I said. "We really don't have enough evidence to really stop production, but I believe in my heart Gregory is doing the right thing to stop until something happened either way."

Mr. Da's driver was waiting to take us to UCPC as we walked out of the hotel on time at 9:30 a.m. The Tuesday morning traffic was normal, and we arrived at the complex at 10:00 a.m. Our driver motioned for us to follow him into the main office and to Ms. Xu's office on the second floor.

"Good morning, Mr. Rick and Ms. Terri. It is good to see you again. I forgot to ask yesterday. I hope my suggestion of Sunday brunch was acceptable?" she said, waving and speaking Chinese to the driver who was standing in the doorway. He bowed and returned to his van to await his next assignment.

"Good morning to you, Ms. Xu," Terri said in response. "Yes, we completely enjoyed the brunch. Thank you for asking."

"I have arranged for you to meet with Mr. Hong, the welder for the assembly line, as you asked. He will meet us in the office of the plant manager."

"Has someone new been appointed plant manager?" I asked, remembering how Mr. Da had fired the plant manager who was on duty when the murder occurred.

"Yes, Mr. Ling is new plant manager," Ms. Xu said.

"And what about the names of the guards and what they had to say about Mr. Alworth leaving the day of the murder?" I asked.

"Oh yes, the guards," Ms. Xu said. "Here are their names and what they said when Mr. Da talked to them."

She handed me a folder with three sheets of paper. I didn't want to take time at that moment to read them, but I thanked her for the information and promised I would read the report prior to our leaving and ask any questions I may have. She nodded her agreement.

Mr. Hong was sitting in the plant manager's office, waiting for us when we arrived. It was obvious he was almost terrified at the thought of talking to us. I tried to make him as comfortable as possible by once again informing him he was not in trouble or someone we felt could have been involved in the murder of Mr. Zhu.

Ms. Xu translated my comments, and Mr. Hong's face relaxed and he nodded to both of us with a small smile of relief.

"Mr. Hong, I am going to ask you several questions regarding what you may have heard or seen in the area of the QC lab on the day of the murder, okay?" I said.

Mr. Hong understood no English and sat looking like a blank sheet of paper until Ms. Xu began to translate. *"Li Cha Xian Sheng Shuo Ta Jiang Wen Ni Ji Ge You Guan Mou Sha Dang Tian De wen, Hoa Ma?"*

Mr. Hong said, *"Wo Jiang Jinke Neug Hui Da."*

Ms. Xu translated, "He says he will answer the best he can."

216

"Okay, please ask him if he was working on the assembly line all day the day of the murder, and did he see Mr. Alworth come to the QC lab and talk to Mr. Zhu?" I said.

Ms. Xu gave Mr. Hong a very stern look and translated, "*Li Cha Xian Sheng Xiang Zhi Dao Shi Fou Ni Zai An Fa Dang Ri Zai Zhuang Pei Xian Shang gong Zou Qie Kan Jiao A Er Wo fu xiamg Sheng Jin Ban gong Shi He Zhu xian Sheng Jiao Tan.*"

Mr. Hong said, "*Mei Dang A Er Wo Fu Xiang Sheng Lai Zhe Wo De Gong Zuo Tai Lai Jian Cha Wo De Gong Zuo. Na Yi Li Ta Ye Shi Ru Ci Wo Kan Jian Ta He Zhu Xian Sheng Tan Hua Dan Wo Mei Ting Jian Ta Men Shou Shen Me.*"

Ms. Xu translated, "Mr. Hong says he always see Mr. Alworth when he come here. Mr. Alworth comes to his station to inspect Mr. Hong's work. He says he did see Mr. Alworth talk to Mr. Zhu, but not hear what they say."

"So, you did not hear Mr. Alworth and Mr. Zhu argue?" I asked.

Ms. Xu asked Mr. Hong, "*Ni ting Jian Zhu Xian Sheng He A Er Wo Fu Xian Sheng Chao Le Ma?*"

Mr. Hong shook his head no.

"Then why," I asked, "did you act so nervous when Mr. Jianguo and Ms. Terri asked you about Mr. Alworth?"

Ms. Xu asked, "*Dang Jian Guo Xian Sheng Hea Er Wo Fu Tai Xing Ni Xun Wen A Er Wo Fu Xian Sheng Shi Ni Wei He Jin Zhang Ne?*"

Mr. Hong shifted in his chair, hung his head and mumbled, "*A Er Wo Fu Xian Sheng Bu Xi Huan Wo. Ta Zong Bu Xi Huan Wo De Gong Zuo Er Qie You shi Xing Wo Pao Xiao. Wo Hen Hai Pa Ta.*"

Ms. Xu said, 'Mr. Hong say he does not like Mr. Alworth. He says Mr. Alworth does not like his work and sometimes yells at him. Mr. Hong afraid of Mr. Alworth."

Terri said, "Then that's why you acted so nervous? Do the people on the assembly line know about this?"

Ms. Xu translated, "*Zhe Jiu Shi wei He Ni Biao Xian Ru Ci Jin Zang Bing Qie Zhung Pei Xian Shang De Yaun Gong Zhi Dao Ma?*"

Mr. Hong replied, "*Shi, Ta Men Ye zhi Dao Bing Qie Ta Men Ye Hai Pa A Er Wo Fu Xian Sheng.*"

Ms. Xu said, "Yes, Mr. Hong says others know and they also afraid of Mr. Alworth."

"Was Mr. Hong afraid if Mr. Alworth did commit the murder and thought he heard or saw something, Mr. Alworth would hurt Mr. Hong?" I asked.

Ms. Xu asked Mr. Hong, "*Ni Shi Fou Hai Pa Ru Guoa Er Wo Fu Xian Sheng Jin Xing Le MoU Sha Qie Ta Ren Wei Hou Ta Hui Shang Hai Ni?*"

Mr. Hong answered, "Shi, shi."

I looked at Terri, and we silently agreed there was nothing else we needed from Mr. Hong. We thanked him and apologized for causing him to have such concern, but asked if he thought of anything that may help us to contact Ms. Xu. Ms. Xu translated my comments to Mr. Hong, who smiled and said yes.

Terri and I spent the next few minutes reading the report on what happened with the guards the day John left the plant. It was clear none of the guards had been on the job of checking those that left the plant that particular time that day. They said they were discussing issues surrounding the change to the daily visitors' logs. The human resources manager had changed the form, and they said they were discussing how to complete it each time someone arrived and left the plant. Prior to the change, no signature was required for someone to leave the plant.

We thanked Ms. Xu and began to walk back down the hall to the courtyard to meet our driver. We were having no luck today. Mr. Hong's words to us regarding John and his sandpaper approach to the assembly line team were not helpful, and I hoped Chief Inspector Ji would not hear the same thing from Mr. Hong.

Terri stopped suddenly and asked, "Why was John even talking to Mr. Hong or the assembly line people? They don't work on products for WMT."

"Damn, that's a good question. I never thought of that."

Ms. Xu was leaving the main office building conference room. Terri said, "Ms. Xu, why would Mr. Alworth talk to anyone on the assembly line, let alone criticize them?"

"Mr. Da really have faith in Mr. John and his quality ability. He tell Mr. John when he is at UCPC to please look at what Mr. Da's people do and let them know if they do wrong, then tell Mr. Da."

"Oh," I said, "thanks, Ms. Xu."

Ms. Xu walked with us to the courtyard, and our driver was standing ready to drive us back to the Grand. It was only 10:00 a.m., but the heat was already becoming uncomfortable. Just as we were about to enter the van, the young assistant ran from the main office building, shouting to Ms. Xu.

Ms. Xu turned and spoke in Chinese to her. The girl stopped shouting and running, but continued to hurry to Ms. Xu's side. They conversed in Chinese for several minutes, and it was obvious the young assistant was agitated about something.

Finally, Ms. Xu turned to Terri and I and said, "It seems one of our women workers come forward and told Mr. Da she saw what she believed was an American leaving through the workers' gate at back of the property on the afternoon of murder."

"Is she here now? Can we talk to her?" I asked Ms. Xu. I knew my voice was trembling. This could be the big break in the case we needed. If the female worker was correct and John was telling the truth about which gate he left the property from, then he was innocent, and we had our proof for Chief Inspector Ji.

"Yes, worker is waiting for us in conference room. Mr. Da is waiting there with her."

We quickly made our way to the conference room. The woman was sitting next to Mr. Da, and you could tell immediately she was scared. She was small with very short black hair with round glasses with no frame. She was dressed in casual clothes, which meant she had not been working that day because the first shift was still on the clock.

We took seats across from her and Mr. Da while Ms. Xu sat next to her. Mr. Da spoke to Ms. Xu for several minutes. Sometimes, he spoke with a stern voice, pointing at the worker and Terri and me. Ms. Xu nodded and also spent several minutes talking to her.

Ms. Xu look at us and began. "This is Ms. Wong. She and her husband work in different areas of plant and different shifts. She is home when he working most of time. On day of murder, she finished work and was in apartment that overlooks the back gate. She was afraid to tell anyone about what she saw, so said nothing. Today, she and her husband did not work and were talking about murder. She told him what she saw, and he told her she had to tell Mr. Da. She was afraid, but agreed.

"She tell Mr. Da, and he have assistant come get us. Mr. Da know you want to talk to her. Mr. Da says she speak no English and is afraid of you, so I will ask questions. Please write down what first questions you have, and I will ask. After first questions answered, then you can write more."

I said Terri and I understood the situation and asked Ms. Xu to first tell Ms. Wong we were very happy she came forward, and we would never do anything to hurt her—to please tell her she was our friend.

It took less than five minutes for Terri and me to write down our questions and hand them to Ms. Xu.

Ms. Xu began asking the first question and replied back to us after Ms. Wong had answered,

"She say she was looking out window thinking what she could want for dinner when she saw this man walk, almost run, out back gate."

"Ms. Xu, is the back gate manned by guards like the front entry gate?" I asked.

Ms. Xu answered, "No, it is not. Mr. Da never thought it was needed; only workers use gate."

Ms. Xu asked the second question and Ms. Wong replied.

Ms. Xu said, "She say she could not tell how big he was, but his hair was brown and short. She tell me all Americans look alike to her, but she know it was an American."

Ms. Wong's reply to the third question was better. She said he was wearing a suit with a red tie. She believed the suit was blue or black. She told Ms. Xu she may have seen the man in the plant before, but was not sure.

Ms. Xu asked our last question.

"She say she does not believe anyone else saw him. Because no one else talk to her about it."

It's funny; gossip is the same all over the world. Too bad someone else had not come forward. Maybe someone else had seen the American leave and would say something now that Ms. Wong had said something.

Ms. Xu started to say something to us when Mr. Da interrupted and talked to Ms. Xu in Chinese. Ms. Xu nodded in agreement and said to us, "Mr. Da remembers Mr. Dureno had been in the plant with a blue suit and red tie before, and Mr. Dureno had short brown hair."

"Dureno!" I asked Terri, "Didn't John say Dureno was in the plant the day of the murder?"

"Yes, he did, but I don't remember the time," she replied.

We thanked Ms. Wong, Ms. Xu, and Mr. Da, telling them they had been extremely helpful. We both bowed to Ms. Wong as she left the conference room and thanked her profusely. She may not have understood English, but she certainly understood our gratitude from our actions.

Ms. Xu excused herself after Mr. Da and Ms. Wong left, telling us she was going to check the last time UCPCs records showed Mr. Dureno had been in the plant. She returned in a few minutes with the daily logs from the front gate. She checked the traffic into and out of the plant on May 5, but Dureno's name was not there. We did confirm John had checked in around 9:00 in the morning, as he had told us.

It appeared the last entry with Dureno's name was on March 29.

There was nothing more to do here, so we once again thanked Ms. Xu as we walked to the main courtyard. I promised to call her with more information after I had called or met with Chief Inspector Ji. I knew he would also question Ms. Wong. I thought our best approach was to tell Chief Inspector Ji what we knew about Dennis Dureno and to ask that we be involved in any questioning of him.

I called Chief Inspector Ji on his cell, but there was no answer. I left a short message explaining the reason for the call and asked him to return my call as soon as he picked it up. My call to the main Changzhou police station told me Chief Inspector Ji was not working that day, but could be reached by cell phone.

We were about ten minutes from the hotel when Chief Inspector Ji returned my call. He was shopping with his wife and daughter, but arranged for us to meet him at the station in a half hour. I handed the phone to the driver, and Inspector Ji told him where to drop us because the driver changed directions, and we arrived at the station in about ten minutes. We thanked the driver, checked in with the desk sergeant, telling him Chief Inspector Ji was on the way to the station, and we would wait for him in the lobby area.

Chief Inspector Ji must have been pretty close to the station because he arrived in less than fifteen minutes. We said our hellos and followed him to his office.

"Now, what's this about someone seeing an American leave the plant through the worker gate on the day of the murder?" he asked.

I began, "Terri and I were at the plant this morning to re-interview a Mr. Hong, who is the welder on the assembly line that sits just outside the QC lab. Sun and Terri had interviewed him earlier in the week and felt he was very nervous and not telling them the whole truth. We decided to return to the plant and talk to him one more time. During the interview,

Mr. Hong answered our question about his nervousness and our feeling he was not telling us the whole truth. His answers told us why, but it has nothing to do materially with the case.

"As we were leaving, one of Mr. Da's assistants spoke to Ms. Xu, telling her Mr. Da was in the conference room with someone who says she saw an American leave the plant through the worker gate around the time of the murder."

Terri said, "Rick and I joined Mr. Da in the conference room, along with Ms. Xu. A Ms. Wong, who works in the plant and lives onsite in one of the company apartments, explained to us that she was at home after working first shift, looking out the window, when she noticed what she believed was an American walking quickly out the back gate.

"Based on her partial description, we identified the person as Dennis Dureno, marketing and sales VP of WMT. We checked the visitors' logs and found the last time Mr. Dureno was officially in the plant was Tuesday, March 29," Terri finished.

"We decided the best thing we could do was to share this information with you and ask if we can be part of the follow-up," I said.

"First, I thank you for bringing this information to me and not trying to solve this case yourselves. Second, yes, you may be part of the interrogation process after we locate this Mr. Dennis Dureno," Inspector Ji said. "Do you have any idea where he might be? I assume he is still in China?"

"I talked to Mr. Brightson Friday evening, Changzhou time. From our conversation, I believe Mr. Dureno is still in China, probably in Changzhou. Mr. Brightson had recently talked to him about his visits to government officials in the Changzhou area."

"Do you know which hotel he stays in while he is here?" Chief Inspector Ji asked.

"No, I am sorry I do not; however, I believe Mr. Brightson will know," I said.

"Let me see: it is 10:00 p.m. on Tuesday night in Charlotte, right?" Ji said.

I nodded that I agreed.

"Will you please call Mr. Brightson and ask him that question?"

"I would be glad to. We were going to call him anyway to tell him what we learned," I said.

Gregory's cell rang only a few times before he answered. It was obvious he was awake and involved in something strenuous, because he was somewhat short of breath.

"Gregory, it's Rick. I'm with Terri and Chief Inspector Ji, and we have some potentially bad news to share. It's a long story that I won't go into right now, but it appears a female employee of UCPC evidently saw Dennis Dureno hastily leaving the plant through the back gate around the time of the murder.

"She has given us enough of a description for us to believe her," I said.

I could tell I now had Gregory's full attention. He was, to say the least, flabbergasted at my news. He kept saying throughout the conversation he just could not believe Dennis would do such a thing. He just kept asking me what motive Dennis would have to kill Mr. Zhu.

"Gregory, I am not sure why he might have done it or if he even did do it, but if he was seen leaving the plant around the time of the murder, we need to talk to him and get his side of the story," I said. "Do you know if he is still in China, particularly in Changzhou, and where he stays while he is here?"

Gregory confirmed Dureno was still in China and believed he stayed in the Jin Jiang International.

We spent the next five minutes discussing what we were going to do next. Gregory asked if he needed to return to China, but I told him there was nothing he could contribute at this time. I promised to call frequently to keep him in the loop.

Chief Inspector Ji was on the phone immediately upon hearing the name of the hotel, asking if Mr. Dureno was registered. He was, but the desk clerk did not know if he was in his room.

"Normally, I would send a team to bring him into the station, but in this case, what do the two of you think?" the chief inspector asked.

I replied, "We met Dennis when we came to China, and I, for one, really don't know how he would react if I called him, and he was the killer. I suspect he would take off." Terri said she agreed with me and recommended Chief Inspector Ji send his team to bring Dureno to the station.

It was 2:00 p.m. when Inspector Ji made one call and a team of four police officers arrived, led by a senior inspector of police. Chief Inspector Ji gave the team instructions where to find Mr. Dureno, while Terri and I provided a description.

We waited in the chief inspector's office while the team was working to locate and bring Dureno back to the station. Our wait wasn't a long one. Dureno looked very confused and aggravated when he was escorted into Chief Inspector Ji's office. He was probably more surprised to see Terri and me there.

"What the hell is going on here, and why am I treated like an escaped convict!" Dureno said. "Does this have anything to do with John Alworth, and why are they here?" he asked, pointing to Terri and me.

Chief Inspector Ji dismissed the team with a nod of thanks and turned to Dureno. "Mr. Dureno, I apologize for taking this precaution regarding my need to talk to you about the murder of Mr. Zhu; however, one cannot be too cautious in these matters. Please, have a seat," Ji said, pointing to the conference table. "You may join us as well," he said, pointing to us and the conference table.

Dureno took a seat on the end of the table, and Chief Inspector Ji sat in the first seat on his right, closest to the door. Terri and I sat further down to stay somewhat remote.

Dureno said, "You really don't think I had anything to do with this, do you? Why would you even think it could be me? I really didn't know the guy. I met Mr. Zhu a few times, but never worked with him. What are you trying to do, pin this on me for a quick answer?"

"Mr. Dureno, before we begin, thank you for cooperating with the officers. Sir, I have not accused you of anything. I merely said I wanted to talk to you."

Before Ji could continue, Dureno said, "I don't believe your guys gave me much choice, did they?"

Chief Inspector Ji answered while trying to avoid a chuckle. "No. I suppose not, but thank you anyway. Mr. Dureno, there has been a development in this case that may involve you, and I would like to hear your side of the story before moving forward. It seems one of the female workers at UCPC was sitting in her apartment on the day of the murder when she saw an American fitting your description leave the plant through the back gate. Can you tell us your whereabouts on May 6?"

Dureno answered, "In the first place, I never knew there was a back gate. I had meetings with the regional development director in the morning until after lunch. The rest of the day, I was involved in something personal on that date."

"Would you like to tell us what was the nature of the personal business and what the times were?" Inspector Ji asked.

"As I just said, I met with the Changzhou regional development director in the morning at the government offices within the regional center. I left there, had lunch near here, and spent the afternoon on a personal matter."

"Let's leave the nature of the personal matter for now," Chief Inspector Ji said. "Can you tell me when you completed the personal matter?"

"I returned to my hotel around seven p.m."

"Can you prove that?"

"No, yes, I mean, I have a cab receipt somewhere, but why should I have to? I did not murder anybody. Hell, I didn't even know the guy! Your witness must have been mistaken about her identification. I didn't even know there was a damn back gate, as I said earlier."

"Might you have that taxi receipt on your person now?" Ji asked.

Dureno spent a few minutes looking in his pockets and asked permission to check his briefcase. He did not find the receipt after a short search. "I guess I must have thrown it away. I normally keep receipts for my expense account, but as I said earlier, this visit was personal."

"Mr. Dureno, where were you between five and five thirty p.m. on May 6?"

"I told you I had a personal appointment around that time."

"Now, Mr. Dureno, I am afraid we must explore the time spent in your personal appointment," Chief Inspector Ji said. "Please tell us the nature of that visit."

"I am afraid I can't do that, Inspector. My visit was very personal and could create an intolerable situation for someone else."

"Mr. Dureno, I am afraid your continued desire to keep your whereabouts from us may also cause an intolerable situation for you."

"As I told you several times, I did not kill Mr. Zhu and have no idea who did, other than John Alworth. If you have nothing else to talk to me about, then release me."

"I can't do that, Mr. Dureno. You must spend at least one night here while I arrange for the witness to make a formal identification," Chief Inspector Ji said.

"Am I being charged with anything? If not, how can you hold me overnight?"

"You are in China, Mr. Dureno. US laws are not the same as laws here."

Chief Inspector Ji walked to his desk and made a telephone call. The senior inspector and one other officer arrived. Chief Inspector Ji gave orders to them, and they asked Dureno to kindly leave with them.

Dureno looked both defeated and deflated as he left with the officers. He shouted to us as he was leaving. "Rick, please call Gregory and tell him what is happening. Maybe he can do something."

I said I would as he left Chief Inspector Ji's office with his guards.

There was nothing more we could do at police headquarters, so we said goodbye to Chief Inspector Ji.

Before leaving, we asked that he please keep us informed as to what was happening with Dennis. Chief Inspector Ji agreed to call us later that evening with the time and date of the eyewitness viewing of Mr. Dureno, but he said he planned to be ready to do so by noon the next day.

We told Chief Inspector Ji we were going to call both Mr. Brightson and Mr. Jianguo. We felt sure one or both would be calling to discuss the day's events regarding Mr. Dureno. Chief Inspector Ji suggested they call him on his cell phone, because he was leaving and returning home to finish his day off.

Fortunately, a cab was sitting empty just outside the police station. I gave him the address of the Grand. The taxi driver nodded he understood and pulled away from the curb into city traffic heading for the hotel.

"I have a real hard time believing Dennis Dureno is our killer," Terri said. "What could he possibly gain from Zhu being dead unless he knew about the eighth life test? I don't believe he or anyone else other than John and Mr. Zhu knew. Of course, if he did find out about the test, I can see where that could be a motive. With Zhu out of the way, production goes forward on schedule. Every one of these guys has potentially millions of dollars to be made if this product is as fantastic as they say."

"Hell, thinking along those lines, I can name a few individuals who might have been in a position for significant financial gain wanting Zhu out of the way. I mean, Mr. Da could be a suspect, if you look at it that way. He has not only a significant personal gain, but also a severe loss for his business if the product does not start production or can't be redesigned."

"I agree with you about Dennis not having a motive. John told me directly no one on his side knew of the eighth life test until John called us and eventually Gregory. Gregory certainly could have said something to

Dennis. If he did, he never indicated that to me. I don't believe Zhu shared his information either. If he had, there would have been no reason for him to request a private meeting with John in Shanghai. He certainly would have lost control of the project if he did."

"And what could Dennis be hiding about where he was at the time of the murder?" Terri asked.

"I'm not sure, but it must be something extremely sensitive," I answered.

"Maybe he is our killer, and he is using the personal appointment as a ruse until he can better solidify his story," Terri said.

"You might be right, but I still can't find a motive, assuming he didn't know about the eighth life test," I said.

We reached the hotel, thanked our taxi driver, and rode the elevator to our suite. "I know it is after midnight in Charlotte, but I have to call Gregory. He said to call him anytime if we needed. This is one of those times," I said as I dialed Gregory's cell number.

"Gregory, hi, it's Rick. I sure hate to bother you so late, but this thing with Dennis Dureno is escalating. Chief Inspector Ji had him brought to the police station for questioning after Ms. Wong said the person leaving the back gate was American and partially described Dennis. Dennis will only say that he was involved in a very personal appointment during the time the murder was committed and continues to deny he had anything to do with it."

I continued, "Gregory, do you know anyone Dennis could have been with over here or anywhere Dennis could have been that is so personal he would risk going to jail over it?" I asked.

"Certainly nothing I am privy to. I know there were times when we were in Changzhou together when he seemed to disappear, and I could not locate him, but I always believed it was an in-country cell connection issue," Gregory said.

"Could he possibly have a girlfriend and be afraid his wife would find out and cause a messy scene? I mean, if this product does as well as predicted when it does go into production, he will be a rich man."

"I would find it hard to believe he would be mixed up in something like that. He seems devoted to his wife and his family. There have been times when we were traveling together in China and he has commented on a pretty girl, and I know he does admire Chinese women, but I never saw him do or say anything out of the ordinary. Are you sure there is nothing I can do?"

"I don't think there is anything you can do right at this moment, but Dennis requested I call you, and I have done so. I will call you tomorrow after the eyewitness viewing for identification. In the meantime, if you think of anything, and I do mean anything, please call me. I believe we will need every arrow in our quiver for Dennis to work out of this one. Of course, if he gives Chief Inspector Ji the name of the person he was meeting with, then he has an alibi and is probably not the murderer."

"Where is Dennis now?" Gregory asked.

"Chief Inspector Ji decided to hold him overnight while he arranges for the witness to come to police headquarters to identify Dennis some time tomorrow."

"Anything we can do about that?"

"No, I don't think so," I answered.

"Have you contacted anyone at Changzhou, Huaide, and Zhanglinfang?"

"I called Mr. Jianguo, Sun, and he is calling Chief Inspector Ji to see what he can do to get Dennis released into our custody. He wasn't very optimistic."

"If you need money, call me back, and I will make sure it is wired to you immediately."

"I will, but something tells me Dennis is there for some time or at the very least until the witness has been at the police station," I said.

I hung up the telephone and sprawled across the bed. It was the first time that day I was able to unwind my mind from the events of the day and concentrate on where we were on the case and examine what we needed to do next.

Terri interrupted my thoughts. "Rick, do you realize it is almost six and we have not eaten since breakfast? Aren't you getting hungry?"

"It is? I have been so wrapped up in everything I forgot about food. I know that's hard to believe coming from me!"

"I know," Terri answered. "You are my clock when it comes to food. Would you like something from room service or what?"

"Room service will be fine. I just thought of something, and I need to call Chief Inspector Ji. Order something we both like. I'll be all right with it."

"You know I don't like doing that. You promise you will eat whatever I order and not complain?"

I replied while dialing Inspector Ji, "Okay, okay with me. Chief Inspector Ji, it's Rick Watson. I hate to bother you, but I need to talk to John Alworth, either in person or on the telephone. I have a couple of questions I must ask him."

Chief Inspector Ji asked if I was going to share both the questions and John's answers. I agreed to do both, and Chief Inspector Ji said someone would call me within the hour.

Terri selected club sandwiches with a salad and fruit for both of us for our very late lunch or early dinner.

Our order arrived in less than fifteen minutes, and we should have known a club sandwich in China was completely different than a US special. The sandwich was constructed like a US sandwich using white bread; however, the contents were far from a US club. The sandwich contained a thick slice of processed ham, Swiss cheese, fried egg accompanied by lettuce, and very unripe tomato slices. All of this was topped with mustard and ketchup.

At least the salad and fruit were acceptable. We salvaged what we could from the sandwich and enjoyed the salad and fruit.

The telephone rang at 7:30 that evening. It was an officer from the main police station advising me that Chief Inspector Ji had requested to have Mr. John Alworth call this number and a Mr. Watson. Was he speaking to Mr. Watson, he asked?

"Yes, this is Rick Watson, and I did request Chief Inspector Ji call the station and arrange for me to talk to Mr. Alworth."

The voice at the other end said, "Just a moment, please."

"Rick, this is John." His voice was booming through the telephone lines. "What's going on? I understand Dennis is in jail also. Is that true, and why the hell is he here?"

"Long story, but it seems one of the UCPC female workers was in her apartment the day of the murder looking out her window when she saw someone she described as an American leaving in a hurry through the back gate the workers use. She partially identified that American as Dennis Dureno."

"Dureno? You mean she identified Dureno?"

"Not completely," I answered. "Her partial description could fit Dennis. Chief Inspector Ji is holding him tonight there with you until he can arrange to have the eye witness brought to the station tomorrow to identify Dennis in person."

"Do you really think Dennis would do something like that? I mean, he isn't the type to kill somebody. Screw you in a business deal maybe, but murder? No way! Besides, what motive would he have?"

"That's actually why I called," I replied. "I need to ask you two questions, okay?"

"Shoot," John said.

"Did you say anything to Dennis about the eighth life test and what it said?" I asked.

"Hell no, I never told anybody about the life test and had no plans to do so until I had the test in my hands and had time to evaluate it. There was no need to panic the entire company."

"Are you positively sure you didn't say something to him, or to anyone else for that matter, maybe some comment in passing that he could have deduced what had happened?" I asked.

"I am positive. Come on, Rick, you know me well enough to know I would never do something like that."

"I know, I know, but I had to ask. Second, do you know if Dennis had any female acquaintances, you know, girlfriends, over here? Anybody he would be so well acquainted with that he would refuse to give Chief Inspector Ji who he was with at the time of the murder."

The other end of the line was silent for several seconds before John answered. "No, I don't think so. He was pretty devoted to his wife and kids. I will admit he did like Chinese women. He was always pointing out those he thought were very attractive, but I never saw him act."

"Johnny, can you think of any reason Dennis may have had that would have led to killing Mr. Zhu?" I asked.

"None. He really didn't even know the guy. He met him a few times on his trips here and, I believe, had been in a couple of meeting with Zhu when he was here."

"Any idea what American would be going out the back gate?" I asked.

"Again, none. Could it have been another one of UCPC's customers?"

"Good thought. I will follow up on that one tomorrow."

"Sorry I wasn't much help."

"That's okay. Hang in there, buddy; I believe we are getting close."

"I sure as hell hope so. I'm sick of this place!"

We said our goodbyes, and I turned to Terri. "You heard the conversation. What do you think?"

"I have always believed John was telling the truth, and it is a good idea to make sure there were no other Americans in the plant on May 6."

"I will do that the first thing in the morning."

We turned on CNN to catch up on the news of the day in North America and both fell asleep with books on our chests.

Chapter 18

Wednesday, May 28, arrived sunny and already getting hot when we awoke around 7:30. We were both starving and decided to take a day off from exercising and headed for the Western-style breakfast room.

Forty-five minutes later, we were condemning ourselves for skipping our morning routine for a fat and calorie breakfast as we rode the elevator to the eleventh floor and our suite, promising ourselves we would not succumb again.

I placed a call to Ms. Xu at UCPC. Fortunately, she was in her office and answered on the second ring. "Ni hao, Ms. Xu," I said. "This is Rick Watson. How are you this morning?"

"Ni hao, Mr. Watson, I am fine, thank you. What may I do for you this morning?"

"Ms. Xu, do you recall if you had any American customers in the plant the day of the murder?" I asked.

"No, Mr. Rick, we do not schedule more than one customer in the plant at a time so we can give customer full attention. Why you ask, Mr. Rick?"

"I just wanted to be sure the American your worker saw leaving by the back gate was not another customer from America."

"I see, no, no other customer," Ms. Xu replied.

I excused myself for interrupting her workday, thanked her, and hung up the telephone on my end.

My next call was to Chief Inspector Ji to confirm the time of the lineup containing Dennis Dureno. Inspector Ji was out of his office, but the officer answering the telephone told me to be at the police station no later than 3:00 p.m. That gave us a few hours to analyze what could be the event or events that led to the killing of Mr. Zhu and, if not John, who could it be.

No matter where our minds wandered, we could not develop a scenario where John or anyone else we knew who we worked with on a daily basis could have committed such a heinous crime.

Terri had plans for the afternoon, so the doorman signaled a waiting taxi when I reached the door leading outside the hotel lobby, and I was on my way to the police station in time for the 3:00 p.m. identification lineup.

The desk sergeant recognized me as I entered the station waiting area. I told him I was there to see Chief Inspector Ji. He made a quick call, and in a few minutes, a young constable arrived and requested I follow him. We took the elevator to the fourth floor where Chief Inspector Ji was waiting for me.

"Good afternoon, Mr. Watson," he said. "I trust your day has been productive."

"Good afternoon, Inspector," I replied. "Let's just say somewhat productive. On a hunch, I called Ms. Xu at UCPC to confirm there were no other Americans in the plant on the day of the murder. She informed me UCPC does not allow more than one customer in the plant at a time, so there were definitely no other Americans in the plant on the day of the murder. I'm not sure how that affects this case, but wanted you to know from me personally."

"It could mean your friend Mr. Alworth is indeed innocent, but I don't buy it. He and Mr. Dureno may be in this thing together," Chief Inspector Ji said.

It was difficult to hide my rising blood pressure as I replied, "I don't really know Mr. Dureno; however, I do know John Alworth, and he is just not capable of doing such a thing."

Chief Inspector Ji smiled, pointed toward the door, suggesting I lead the way. The lineup room was very similar to those I had seen in the United States. There was a small room with several chairs placed in front of a two-way mirror. Behind the mirror was a platform about two feet off the floor and approximately twelve feet long and five feet deep. There was just enough room for the suspects to stand and turn on command.

Ms. Wong was already in the room, sitting in an oak straight back chair positioned in the middle of the room facing the window. She was dressed in what appeared to be her best clothes and was acting extremely nervous. I felt she was also uncomfortable with being inside a police station as much as she feared what could happen if she identified Dennis Dureno.

Chief Inspector Ji introduced everyone in the room to the others there and made a call through the intercom to the officer charged with controlling the lineup.

Six American men filed into the room spreading across the full length of the stage area. Dennis Dureno was in position number four.

Chief Inspector Ji began in Chinese, "Okay, now, Ms. Wong, please look at each man individually, take your time, and please tell us if the man you saw that day is in the lineup. Take your time, don't be nervous, just look closely at each person and let us know if you see the man."

Ms. Wong looked intently from one face to the other with no apparent recognition of any of the individuals in the line. The officer asked if she would like for any of the men to turn right and left. She nodded yes, so he communicated that to the officer inside the room. The officer shouted to each of the men to turn right and left as he passed by.

Ms. Wong was very quiet, and I thought she wasn't going to say anything about any of the men in the lineup until she finally pointed to Dennis Dureno and said, "That one, that one, he could be the man I saw, but I am not really sure. All Americans look alike to me."

Chief Inspector Ji walked to Ms. Wong's side and kneeled down to be at her height. "Ms. Wong, it is very important you be right about anyone you identify. Please take all the time you need and ask any questions you have."

She continued to stare at the six men before her before finally saying she could not be 100 percent sure, but she thought number four was the man. Chief Inspector Ji used the intercom once again to dismiss the men and ordered the ones that were not plants be returned to their cells.

"Well, where do we go from here?" I asked Chief Inspector Ji.

"I plan on talking to Mr. Dureno immediately and confront him with this new information. Maybe he will be more cooperative this time!"

"May I join you?" I asked.

"Of course," he replied.

Chief Inspector Ji led the way to the same interview room in which I last saw John. He made a call, and within five minutes, Dennis Dureno was led into the room by an officer.

"Good afternoon, Mr. Dureno," Ji said. "I have asked Mr. Watson to join us."

"Hey, Dennis," I said.

234

Dennis nodded at each of us and said, "Am I free to go?"

Chief Inspector Ji replied, "I am afraid not, Mr. Dureno. You were identified during the lineup by our eyewitness."

"That's fucking impossible!" Dureno was beginning to become highly agitated. "There was no way she could identify me when I wasn't there!" Dureno shouted at Chief Inspector Ji. "Who is this person? I want to talk to her. Get her in here." Dureno's voice became louder and louder as he spoke.

"Mr. Dureno, that's impossible. Now just calm down and tell us what really happened that day in the plant," Chief Inspector said. "I must warn you, it will be much easier on you if you tell the truth now rather than later."

"How many times do I have to tell you I wasn't near the plant that day, and I really don't know Mr. Zhu. I had nothing to do with whatever happened to him, and you can't prove I did it."

"Mr. Dureno, we have a witness who can place you at the scene at the time of the murder, and we know you have a personal reason to ensure the product begins production on time. I believe if we look further, we will find why you did it. By the way, was John Alworth in on it with you?" Chief Inspector Ji asked.

"In on it with me, I thought you had all the evidence you needed to convict him, why me now?" Dennis asked sarcastically.

Chief Inspector Ji continued, "We do have evidence against Mr. Alworth, and now we know you were also involved. Who pulled the trigger, Mr. Dureno? Did you shoot Mr. Zhu and give the gun back to Mr. Alworth, or did he do it?"

"Jesus Christ, are you nuts!" Dureno said in a wail. "I said I had nothing to do with this, and I was certainly not involved in anything John Alworth may or may not have done. You have got to believe me! Tell him, Rick; you know I had nothing to do with this!"

"I would like to believe that, Dennis, but now I am not certain. All I know for sure is John Alworth did not kill Mr. Zhu."

"This can't be happening to me. How can both of you keep saying these things that are not true!" Dureno's face was contorted as he screamed. It was obvious Dureno was working himself into a panic and a complete denial of anything he may have had to do with the killing. "All I want to do is get out of here!"

"Once again, Mr. Dureno, I am afraid that's not possible," Chief Inspector Ji answered. "I believe you and Mr. Alworth were in on this together, and I will prove it."

Dennis pounded the table in complete frustration after each word to emphasize his statement, "I—did—not—kill—anybody!" His words came from clenched jaws. "Can't you understand that!"

"Officer, please escort Mr. Dureno back to his cell," Chief Inspector Ji said. "And Mr. Dureno, I plan to formally charge you as an accomplice to the murder of Mr. Zhu no later than tomorrow morning. I suggest you engage counsel as quickly as possible."

Dureno gave me a pleading look and, almost begging, said, "Rick, you have got to contact Mr. Jianguo at Changzhou, Huaide, and Zhanglinfang for me." I said I would do that for him as soon as I left the police station.

Dureno hung his head and began to cry as he reached the door that led back to his cell. He acted as if a chill had crossed his back and turned to face both Ji and me. "Okay, okay, you win. I'll tell everything. Why I am not involved in this and why I am not guilty. Just let me sit down and compose myself."

Chief Inspector Ji motioned for the guard to return Dennis to the table. Dennis sat down, stared at the ceiling for a few minutes as if he was coming to grips with something of such magnitude he had to sell himself on the idea of confessing.

He began, "Yes, it was me that ran out the back gate the day of the murder, but I didn't kill Zhu. I was with someone most of the afternoon. I had been in the plant earlier in the day and wanted to talk to John some more about the seventh life test and to see if John was going to stay in Changzhou that night. If he was, I was going to suggest we have dinner somewhere downtown.

"I walked from where I was that afternoon and came into the plant by the back gate. When I meet with my friend, it is only a short walk to the back gate of the plant, so I normally come in that way."

"Which gate do you normally exit from?" Chief inspector Ji asked.

"Always out the front gate."

"Did you sign out the day of the murder?" I asked.

"No, since I didn't sign in, I didn't sign out," Dennis replied.

Chief Inspector Ji said, "Go on."

"I made my way across the campus and into the QC lab. That's where John spends most of his time, and I thought I might catch him there."

"What time was that?" Chief Inspector Ji asked.

"I remember looking at my watch as I rounded the corner of the lab thinking I might have missed him. My watch said five forty-five."

"Go on," Chief Inspector Ji said.

"I opened the door, and before I could call out John's name, I saw a body lying on the floor in a pool of blood. I looked to make sure it wasn't John and recognized Mr. Zhu. I panicked. I ran out of there as fast as I could and headed for the back gate. I was praying all the way I would not see anyone who might recognize me. I went out the back gate because I knew I would have to sign out if I went out the front. So much for not being recognized, eh!"

"Where did you go after you left the plant?" I asked.

"I went straight to the nearest bar and had a drink to steady my nerves."

"And where did you go after that?" I asked.

"I went to my hotel room for the rest of the night."

"Did anyone see you in the plant or at the bar that could verify your story?" Chief Inspector Ji asked.

"No, nobody. I had never been to that particular bar before, and it was full of college students celebrating something. Everyone in the bar was paying attention to them, including the bartender."

"How about the hotel? Anybody see you come in? Did you talk to anyone? How did you get back to the hotel?" Ji continued, "Did you stay in your room the rest of the evening?"

"No, I don't think anybody saw me come into the hotel. At least, no one spoke to me directly. Like I told you before, I took a cab back after leaving the bar, but I have lost the receipt."

"Any bar tab in your pocket?" I asked.

"No, I paid cash."

Chief Inspector Ji said, "Assuming I believe what you just told me is true, you still have no one to back up your story. When you began a few minutes ago, you said you were going to tell us everything. Who were you with that afternoon? We need to contact them and verify your story."

Dennis sat there once again as if he was trying to find a way to not tell the next part of his story. I believe he decided he could not, so he began, "You must keep what I am about to tell you very confidential and promise not to make the name of the person public to anyone, including city officials!"

"I can make no promises to you, Mr. Dureno. I will only promise to be as discreet as possible, but my first obligation is to solve this murder, so who is it?"

Dennis struggled to let the letters escape from his mouth like someone expelling his last breath. "For the past two years, I have been involved with Ms. Marie Dong."

Before Dennis could continue, Chief Inspector Ji whistled under his breath, causing everyone to look in his direction.

Inspector Ji asked, "Did you say Ms. Marie Dong?"

Dennis replied, "Yes, I am afraid so. Now you know why I could not say anything initially."

"Who is Marie Dong?" I asked.

Chief Inspector Ji answered quickly, "She is the first special assistant to the mayor of the city."

"Oh, Jesus Christ, Dennis, why her? Why not someone with less notoriety?" I said.

"It just happened. I met her during my first visit to the mayor's office and couldn't get her out of my mind," Dennis said. "I really tried to stay away from her, but I could not. I asked her out for dinner, she accepted, and we have not been apart since. I spend as much time as possible with her every time I am here."

"Does the mayor know?" Ji asked.

"My God, no!" Dennis said, the panic returning in his voice. "Our business with the city is much too valuable for this to come out."

"I assume she will verify you two were together all afternoon on the day of the murder?"

"I have discussed the situation with her, and yes, she will, but please, be discreet so the mayor doesn't find out."

"No promises, but I will do the best I can."

"Any more questions, Rick?"

"Only one. Dennis, where do you normally meet Ms. Dong, and how long is the walk from there to the plant?"

"We normally meet at her apartment, about a ten-minute walk from here."

Chief Inspector Ji motioned for the guard. "Mr. Dureno, I will be contacting Ms. Dong, and if she verifies your story, you will be free to go."

"Thank God!" Dennis said. "Please, do what you can to keep the mayor from finding out."

"As I said, no promises," Chief Inspector Ji answered.

The guard escorted Dennis from the room and back to his holding cell.

"You know, Chief Inspector, if the young woman will vouch for him, I believe we can trace the walk to the plant from her apartment. If it truly is a ten-minute walk, then Dennis could not have been in the plant when the murder was committed."

"You are correct, Mr. Watson. I plan on calling the young woman immediately and meeting with her to obtain a statement. Would you like for me to call you when I have spoken to her?"

"Yes, I would appreciate it if you would," I replied.

We made our way to the lobby, where I thanked Chief Inspector Ji for allowing me to be a part of the discussion. I left the police station and found a waiting taxi just outside the station walls.

Terri was in the suite when I arrived at 6:00 p.m. She took one look at me, crossed the room, and gave me a big hug. "Rough day?" she asked.

"Not so much that, but I hoped we would find something more to prove John's innocence," I said, returning her hug.

I spent the next hour bringing her up to speed on the day's events.

"I guess Dennis had everybody fooled, didn't he! Everyone we talked to said there was no way he could have been involved with someone here. Just proves once again you never really know what goes on in someone's life," Terri said. "Do you believe him?"

"Yes, I do. If you had seen the anguish he went through before giving Chief Inspector Ji the name, you would agree with me," I said. "I'm famished. What or where's for dinner?" I asked.

"It's late, so how about the Western restaurant in the hotel?"

"That's fine with me," I said. "Chief Inspector Ji is going to call me after he talks with Ms. Dong."

We were into our entrées when my cell rang. It was Chief Inspector Ji.

"Hello, Chief Inspector. Thanks for calling me."

"Mr. Watson, we have talked to the young woman, and she had confirmed his story. I also sent one of my best men out to do the walk from Ms. Dong's apartment to the plant. Mr. Dureno was telling the truth. The walk took ten minutes. That means he could not have killed Mr. Zhu. I also do not believe Mr. Dureno was involved with your Mr. Alworth. We have released him this evening."

"Okay, Chief Inspector," I said. "Again, thank you for calling me and allowing me to work with you on this case. I plan on returning to the plant tomorrow to continue my chase for the eighth life test. I will keep you informed of my progress."

"Goodnight."

We finished our dinner, and neither of us was ready to give up on the day. The lobby bar was a perfect place outside our suite for a nightcap. Terri and I sat there for several minutes after we placed our drink orders without saying anything to each other. Both of us were lost in our own individual thoughts regarding the case, the events of today, and where we would go from here.

I spoke first. "I really thought we had a chance to prove John innocent when Dennis came into the picture. I wasn't sure how he fit, but I was sure we could use his possible involvement as a way to, at least, get John out on bail. That's certainly not going to happen now."

"You know, honey, I feel we are so close, but I am not sure why," Terri said. "Have we explored all the avenues where the eighth life test may be hidden and who hid it?"

"That's where I am going to begin again in the morning. I plan to be at the plant early to continue my search."

"What about Billy? Do you think he knows more than he is telling us?"

"Could be, but I doubt it. He seemed to want to find out what happened as much as I do. I will push him some tomorrow just to see what happens."

Terri said, "It's after ten in the morning in Charlotte. Do you think we should call Gregory and tell him what happened?"

"Oh damn, I completely forgot about Gregory! Maybe Dennis already called him, but I should as well."

I placed a call on my cell to his cell number, and he answered on the second ring. "Hello."

"Hello, Gregory, this is Rick Watson. How are you this morning in Charlotte?" I asked.

"I guess things are fine in Charlotte," he said chuckling. "I'm in New York City, just about to leave for a meeting with some bankers. What's up?"

"Well, first, good luck with the bankers, and second, do you have a few minutes to talk about the outcome of the police arrest of Dennis?" I asked.

"I'll make time. The bankers can wait. What you have to report is more important! What happened?"

I spent the next several minutes giving him a detailed verbal report of the day's events, including everything Dennis said about why he actually was in the plant, finding the body, running out, and the fact that he did have an ironclad alibi. I left out the part about Ms. Dong. I believed it was up to Dennis to talk to Gregory about that. Fortunately, Gregory did not ask about the alibi.

"I guess what you said is good news for Dennis, and I am happy about that, but bad news for John and the production start," Gregory said. "I met with the board on Monday, and they agreed with me that starting production without concrete evidence there is or is not an eighth life test is a bad idea. I would assume that means also we are no further in proving John is innocent and have not found the eighth life test."

"Yes on both counts," I replied. "I plan to return to UCPC tomorrow morning to continue my hunt."

"Rick, as I have said, I really appreciate what you are doing for John and WMT, and I personally plan to keep resisting starting production. I don't know how long I can hold the board and some large investors at bay, but I will do my level best," Gregory said. "Do you believe you are any closer to solving this thing?"

"It's like Terri said to me tonight: she believes we are so close, but can't explain why. I feel the same way. The answer is out there, but we keep missing it."

"I know you and Terri will find the answer. Please keep working on it."

"We will, and thank you for your confidence."

"I really need to go now. You can only keep bankers waiting so long!"

"Okay, Gregory, and thanks for taking the call. We will be in touch again soon."

"Goodbye and good hunting, Rick," Gregory said as he hung up.

"With that," Terri said, "I believe it is my bedtime. How about you?"

"I guess so," I replied. "I am not sure I can go to sleep, but let's go."

"You, my husband, not fall asleep. Ha! Two seconds after your head hits the pillow, you will be out!"

Chapter 19

She must have been right because the next thing I remembered was sunlight in our room and the beginning of Thursday, May 29.

We quickly slid back into our morning routine with a longer run than usual. It felt good to work off some of my frustration. Terri must have felt the same way, because she pushed me the entire run.

We showered and primped for the day. I placed a call to Ms. Xu at UCPC requesting permission to return to the plant and the QC lab to work with Billy. Ms. Xu readily agreed and said she would make sure Billy was available. She also told me Chief Inspector Ji had called Mr. Da very early this morning to tell him about what occurred yesterday with Mr. Dureno.

I asked her if she thought I should also meet with Mr. Da, but she said he was out of the plant for the next three days visiting customers in Northern China. Ms. Xu did say she would ask him if he would like to talk to me when he called in later in the day.

I hung up the telephone when Terri said she was going to call Ann Alworth before her bedtime to make sure she also knew what happened with Dennis Dureno. I agreed and asked her to also inquire when Ann would be returning to China.

The doorman signaled for a taxi just as soon as I exited the revolving door. He gave the driver the address of UCPC, and we pulled into morning traffic. We had only gone about two blocks when my cell phone rang.

"Hello," I answered.

"Listen, buster, once again you are in big trouble!" Terri said in a very stern voice. "You traipsed out of here without so much as a goodbye, a kiss, or even a kiss my ass! You will certainly be foiled again tonight with an attitude like that!"

She was right. I was so wrapped up in planning what I was going to do during the day I completely forgot to say goodbye or kiss her. "Guilty as charged, madam. I humbly apologize and kindly request what it is that I can do to once again win the fair maiden's heart?" I asked.

"Let me see," she replied. "How about roses by the dozens, Snickers by the box, or just get your butt back here early so you can buy me dinner at Zhangshengji, and I promise, there will be no foiling!"

"An early return it is, and I will make reservations for a private room at Zhangshengji's. It's the least I can do regarding the circumstances," I said.

"Nice try. You know there are no private rooms in the Zhangshengji, buster. I want dinner in the main room so other diners will see you grovel!"

"My, you drive a hard bargain, but I am temporarily at your mercy," I said, laughing.

"You bet you are, and I plan on making the most of it!" she replied. "Be careful and get home safe. I am already thinking about what I want to order!"

"That is very cold, my dear. See you this afternoon."

"Bye," Terri replied.

Billy was already at work in the QC lab when I arrived. He looked up as I entered and wished me good morning. I returned his salutation as I was placing my briefcase on an empty desk and began to unload it.

I waited until Billy had completed measuring the part stationed on the CM machine before asking him, "Billy, it has been a few days since we last worked together, and you have had time to think about what has happened. Do you still believe there was an eighth life test?"

"Yes, Mr. Rick, I do," he replied. "I know Mr. Zhu was working on something, but he never tell me."

"How can you be so sure?" I asked.

Billy stopped what he was doing and said, "I overheard Mr. Zhu call WMT engineering and request another water sample just like the other seven."

"He did what!" I was stunned at Billy's comment. "You are telling me Mr. Zhu called WMT engineering and requested another water sample?"

"Correct, Mr. Rick."

"Why didn't you tell me this before, Billy?" I asked.

"I don't know. I was afraid I would be in trouble with Mr. Da if I said something, but I think about what you say and decide to tell you. I know how important start of production is for Mr. Da and whole plant."

"You did the right thing, Billy. Don't you worry about Mr. Da. I will handle that situation if Mr. Da has a problem with anything you said," I said.

"Thank you, Mr. Rick."

"Billy, who did Mr. Zhu talk to in engineering at WMT?" I asked.

Billy replied, "He talk to Mr. Rob Johnston."

"And who is Mr. Johnston?" I asked.

"I am not sure, but all samples we get come from him," Billy replied. "I see his name on paperwork."

"Did you get an eighth sample from WMT and Mr. Johnston?" I asked.

"Yes, sir, we did."

"Billy, my boy, you just made my day!" I said, giving him a hug. Billy looked at me with a shocked look, but laughed at me.

I couldn't wait until the end of the day to tell Terri about Billy's comments, so I stepped outside the plant and called. She answered on the first ring.

"Hello."

"Honey, you are not going to believe what just happened," I said. "Billy told me Mr. Zhu called WMT Engineering and talked to a Rob Johnston, requesting an eighth water sample be flown to UCPC. Billy says they got the sample, and he is sure Mr. Zhu used it for the additional life test. Now we know officially there was an eighth life test performed. Now all we need to do is find it!"

"That's fantastic, honey. Are you going to call Rob Johnston to confirm what Billy told you?" she asked.

"Late tonight. It is too late to call him now, and I don't have a number for him. I could call Gregory for the number, but I can wait until tonight."

"I have news as well," Terri said. "I reached Ann on her cell in Charlotte. She already knew about what happened with Dennis. She seemed very upset that he was not the guilty person. She told me she was returning to China on the early flight tomorrow and would arrive in Shanghai Friday. She told me she was going to stay at the Shangri-La until early next week. I thought that was rather odd. I would have thought she would want to see

John as soon as she was in country. I asked her, and she said she had clients to meet with first."

Terri continued, "Maybe we will learn some more tonight talking to Mr. Johnston. Are you still going to be early?"

"Absolutely," I replied. "See you soon."

"Love you, too, babe," Terri said. "See you tonight."

"I screwed up again, didn't I?"

"You sure did. I will be glad to get you back home so I can retrain you! Bye."

She hung up before I could respond and defend myself.

Billy and I spend the afternoon going over every piece of documentation on all of the life tests. We reconfirmed each one was acceptable and did not find any evidence of the eighth life test. I talked to Ms. Xu about any other place in the plant Mr. Zhu might have stored information. She did remember there was an archive room where some quality documents were stored. Billy's schedule required he complete a series of first articles for another significant customer, so I was on my own in the archive room.

By 3:30, it was obvious there were no current portions of life test data in the archive room. Everything was at least five years old.

I decided it was time to pack up and head for the hotel. There wasn't much I could accomplish at the plant. Once again, I promised Billy I would support him if there was a problem with Mr. Da regarding the water sample for life test number eight.

Ms. Xu had arranged for the driver to deliver me back to the hotel. I thanked her and climbed into the air conditioned van for the ride back to the Grand.

We were barely ahead of the evening rush hour, so the ride to the hotel was only twenty minutes rather than the normal hour. My cell rang during the ride. It was Dennis Dureno.

"Hello, Rick, thanks for taking my call. I wanted to call you to thank you for helping me yesterday with Chief Inspector Ji. I know that had to be a problem for you, and you're trying to get John out of jail."

"No problem for me, Dennis," I said. "Both Terri and I never believed you were involved in the murder in the first place. Were you able to keep you situation under control?"

"Yes, thank God, I was, and I have learned my lesson. There will be no more situations like this one. It took yesterday for me to realize what I have at home and how much I didn't want to lose it."

"I am glad to hear that on both counts," I said. "Actually, I was going to call you. Do you know a Mr. Rob Johnston at WMT?"

"Sure. He is the engineer in charge of the Master Clear Water Filtration System. Why are you asking?"

"It seems Mr. Zhu requested Mr. Johnston send UCPC an eighth water sample, and that was the sample he used to run the extended life test."

"You're kidding, so there really was an eighth life test," Dennis said.

"It appears so," I replied. "I plan on calling Mr. Johnston this evening to discuss why and when he sent the sample. Would you like to join me on the call? You can introduce me to him. I believe you all have a conference line number. Am I right?"

"We do. I will set the call up with a call-in number and password for each one of us. What time would you like the call?"

"How about ten p.m. our time or ten a.m. his time."

"Works for me. Would you like to include a topic for discussion?" Dennis asked.

"I think we can say something about production of the new product. That will whet his appetite," I replied.

"Sounds good to me. I will send you an e-mail this afternoon with the numbers and confirmation for the meeting."

"Thanks, Dennis, I will talk to you tonight, and by the way, you made the right decision."

"You bet I did, and thanks."

A note attached to my swimsuit sat on the dining table informing me Terri was in the pool and to join her. It didn't take me long to change and take the elevator to the third floor pool. There was no surprising her this time. She saw me come in the door.

"Hi, honey, come on in; the water is warm and relaxing."

I dropped my things next to hers on a round table situated near the door leading to the pool and slid into the warm water. I had to admit it did feel good after the day I had.

We swam and lounged around until six, and then went to the room to shower and change for dinner. Dennis's e-mail was in my inbox when I checked while Terri was in the shower. Rob Johnston had confirmed

10:00 in the morning his time for the call. Dennis had included the dial-in number and password.

"What time did you say your call was tonight?" Terri asked.

"Ten o'clock our time, ten in the morning their time," I responded.

"Good, that gives us ample time for dinner, and you should still be fresh after your call."

"Don't you worry about me," I said. "I will be ready for a little foiling later tonight."

Terri laughed and shoved me in the direction of the shower. We were out the door of the Grand at 8:00 with enough time to enjoy a leisurely dinner and still be back in time for the call to Charlotte.

Once again, we marveled at the quality and creativity of the food and the service at Zhangshengji. We were truly going to miss it when we returned to the United States. Terri ordered a curry shrimp dish that contained the largest shrimp either of us had ever seen. I stuck with the more traditional full fish dinner, Beijing style. We both selected sizzling rice soup as our starter. The vegetables were perfectly seasoned and crunchy, which is not always the case in China.

At 10:00 p.m., I dialed the number for the conference call and was informed the moderator had not yet joined the meeting. Music played for a few minutes until there was a beep and Dennis Dureno came on the line. The line beeped one more time, an indication there were three people on the call.

"Good evening, guys," Dennis started. "I should have Rob Johnston in Charlotte and Rick Watson in Changzhou. Are you both there?"

Rob and I answered in the affirmative.

"Okay, so, let's begin," Dennis said. "Rick, will you please introduce yourself to Rob, and Rob, please do the same to Rick."

I gave Rob a brief outline of my and Terri's background and the reason we were in China trying to prove John did not murder Mr. Zhu. I brought him up to date on my work with the seven life tests and my newfound knowledge of his shipment of a water sample for number eight.

Rob explained that he was an original employee of the company hired initially as an engineer. He had worked his way up to vice president of engineering and had been selected by Mr. Brightson as the project leader for the development and production of the Master Clear Water Filtration System. He said that one of his main jobs as the project leader was to manage the integration of the software and hardware and to support John Alworth in the quality approval process.

"Thank you both for the synopses of your careers and why each of you are involved with the new product," Dennis said, once again assuming control of the conference call. "Rick, I believe you have a few questions for Rob."

"I do, and thanks, Rob for talking with me. I understand that a few months ago, I am sorry I don't know the exact date, you shipped an eighth water sample to UCPC at the request of Mr. Zhu," I said.

"Yes, I did. He called one day and said he had some equipment problems during the testing and needed one more sample to complete the seven life tests. Mr. Zhu said everything was testing out as expected, and he believed John and I would be pleased when we saw the results. I did not think any more about it. I had one of my engineers obtain a water sample from a predetermined backup city, and we shipped it to him."

I asked, "Did you talk to John about the request?"

"No, I am now not sure why I didn't, but I guess I thought it would be okay to ship the eighth sample," Rob replied. "I really screwed up, didn't I?"

"I wouldn't say you screwed up, but if you had talked to John before you sent the sample, I am sure he would have engaged Mr. Zhu to find out why. When John arrived in Changzhou, Mr. Zhu confronted him with information on that eighth sample. According to Zhu, he had run an extended life test and found a fatal flaw in the design."

Rob replied, "Mr. Brightson told me about what happened when John arrived there, the extended life test, the possible failure, and Mr. Zhu wanting to meet John in Shanghai with the evidence. I understand you have not found a trace of it. You don't really believe John had anything to do with his murder, do you?"

"If I did, I wouldn't still be here! I have known John for too many years. He isn't the kind of person who could take another person's life."

Dennis interjected a question: "Rob, I know you and your team have been working overtime trying to duplicate, with little firsthand knowledge, what Mr. Zhu might have been talking about. Have you made any progress?"

"Absolutely none. We have run extended life tests lasting various times using various samples and have not found anything like what Mr. Zhu told

John he was able to produce. Rick, do you believe there is an eighth life test?"

"I wasn't sure initially. I wanted to believe John, but after all the work I have done to find it, I was beginning to have just a smidgen of doubt. Billy's confession earlier today clinched it for me. We now know there is such a test, and somehow, we will find it. Rob, please permit me one more question," I said.

"Okay."

"Is there any doubt control sample number eight met specifications?"

Rob replied, "I am positive that sample was to specifications. All drinking water in the United States is controlled by federal standards. Each water treatment system may be slightly different from others, but the sample was not out of specifications."

He continued, "I wish I was there to assist you or at least give you moral support. Let me know if there is anything I can do from here."

"I will," I said. "I need a little time to think where to go from here. I may be back to you sooner than you think."

"Again, I will do what I can. I feel somewhat foolish to have let this happen."

"I appreciate it, Rob."

"I guess that wraps up what we sat out to accomplish. Thanks to both of you," Dennis said. Rob and I said goodbye, and Dennis disconnected the call.

It was 11:00 when the conference call was completed, and suddenly, I was very tired. A large yawn escaped from my mouth. Terri walked back into the room just as the yawn reached its peak. She laughed and said, "Oh no you don't, buster. I have waited all day to have some fun. If you think you are going to get out of it, surprise, you're not! You promised me!" She pushed me onto the bed before I could react and what followed woke my senses up. In fact, I had a hard time going to sleep afterward.

Chapter 20

We were greeted by another sunny morning on Friday, May 30. I believe subconsciously we never fully closed our bedroom curtains at night because we enjoyed the morning sun waking us.

The day started with our normal routine of stretching, exercise, sauna or steam at the gym, and a light breakfast in the executive lounge. I hadn't fully developed my schedule for the day, but thought I would call Chief Inspector Ji and ask permission to have a look at Mr. Zhu's apartment. Maybe a search would not be successful; the apartment was the last piece of the puzzle of which I had no intimate knowledge.

I was also going to call Rob Johnston later that night and request he send another water sample to UCPC. If I couldn't find the eighth life test, maybe I could recreate it. I knew Mr. Da would be agreeable to providing the resources and equipment for the test.

Terri's cell phone rang, dragging me from my thoughts to reality.

"Hello," Terri answered, giving me the "who could this be?" look. "Good morning, Ann. Are you back in Shanghai?"

"Yes, my plane landed early, and I am on the way to Changzhou. I haven't talked to either one of you lately, but Gregory has kept me up to date on any progress. That was sure a cock-up over Dennis, wasn't it?"

Terri replied, "I believe it was, but everything is okay now."

"Maybe okay for Dennis, but my husband is still in jail. I plan on seeing him this afternoon. Do you need me to do anything when I see him?"

"Honey, do you need anything from Ann when she sees John this afternoon?" Terri asked.

"No, not that I can think of," I replied. "Maybe she can bring a cake with a small explosive device planted inside to help John escape!"

Terri turned her nose up at that remark and said to Ann, "No, we are okay here."

They spent the next few minutes discussing where we were on proving John innocent and the possibility of doing a ninth life test. Ann voice belied her outside calm. There was no doubt she was anxious about freeing John from the jail at police headquarters.

"How long are you going to be in China this time?" Terri asked Ann.

"I plan on being in Changzhou to see John and Mr. Jianguo and then going back to Shanghai. I have arranged a meeting with Mr. Jianguo for tomorrow morning to see if there is anything else we can do to force the police to release John on bail. I can't believe there isn't something we can do to make that happen."

"Good luck. Maybe Mr. Jianguo has had enough time to develop a game plan on John's release, but Rick and I doubt if Chief Inspector Ji will ever agree."

"You are probably right, but I have to keep trying if only for my peace of mind," Ann replied. "I mean, they have to move his case along sooner or later!"

"If you wouldn't mind, would you please call us after you have met with both John and Mr. Jianguo?" Terri asked.

"Absolutely. I will call you later tonight," Ann said as she was hanging up.

Chief Inspector Ji was in his office and agreeable to my doing a look through of Mr. Zhu's apartment. He said Constable Ming would meet me there at 1:00 p.m. I responded that I did not know where the apartment was, so Chief Inspector Ji changed the plan. The constable would be waiting in front of the Grand at 1:00 p.m. instead.

I exited the hotel at 1:00 p.m. to find Constable Ming standing beside his patrol car smoking a cigarette. "Mr. Watson?" he asked. "I am Constable Ming. Chief Inspector Ji say I drive you to Mr. Zhu's apartment, stay with you, and drive you back. Okay?"

"Ni hao, and xie xie, Constable. Are you ready to go?" I asked.

"Shi, we go now," he answered.

Mr. Zhu's apartment was about a mile west of UCPC in an area of eleven high-rise apartment buildings that each were at least thirty stories tall.

Constable Ming pointed out the building sitting in the rear of the complex facing a small stream. The building, like all the others, was made of concrete in a triangular shape. The architecture was quite simple with defined angles and flat surfaces. Each apartment had ample windows for viewing a combination of country side and city. Mr. Zhu's apartment was on the twentieth floor on the left side of the triangle. We took the main elevator located in the building that acted as the base for the triangle. Mr. Zhu's apartment was a left turn out of the elevator. His apartment was number 2005 once you reached the left side of the triangle. The door was still identified as a crime scene. Constable Ming removed the obstruction and unlocked the door.

The apartment was small by American standards and was only occupied by Mr. Zhu. The furniture was comfortable looking but aged. The living room contained a small brown tweed couch with a small accenting brown loveseat. There was an ottoman in the center of the L-shaped arrangement that had obviously been used only as a foot rest. The entire apartment was devoid of personal items. There were no family pictures and no pictures of Mr. Zhu himself.

There was a small eat-in kitchen with a two-person table. The table was chrome-plated metal with a red laminate top. The two chairs were red simulated leather with chrome legs and base.

The one bedroom held a double bed with a powder blue comforter and powder blue pillow covers. I wondered if he could be a Tarheels fan. You would be surprised at the number of NCAA basketball fans there were in China. I really did not know Mr. Zhu personally prior to his death, but I seriously doubted he selected the bedspread and pillows. He had to have had help. A worn oak desk and straight-backed chair had served as an office when Mr. Zhu worked there.

One bathroom was situated across from the bedroom, and there was no balcony leading out from either the living room or the bedroom areas. The windows, however, did unlock and slid open to the right. I was surprised by that. Code in the United States would not tolerate a window at that height in a building to open.

Everything throughout the apartment was in the place you would expect it to be. All the rugs, window treatments, and dishes looked like the comforter in the bed room. Dishes from a meal prior to his death were still in the kitchen sink.

Someone had to have helped Mr. Zhu decorate in the manner it was decorated, most likely a female.

I asked Constable Ming if he knew if Mr. Zhu had any family. He replied Mr. Zhu's wife had died several years ago, and he had a daughter who was in her twenties. Constable Ming believed she was unmarried and worked in Shanghai.

The afternoon was both frustrating and uneventful. I searched every room from top to bottom, looking for any place where Mr. Zhu could have hidden the eighth life test data. I looked behind pictures, in tops of closets, under the sinks in the kitchen and bathroom, and for possible trapdoors on the floor. I tried to think of every police show I had seen on television and how the actors conducted a search, but nothing worked. I found nothing to make me suspect there was something of value in the apartment.

Constable Ming and I closed the apartment door right at five in the evening and headed for the hotel. I thanked him for the ride and his assistance. When we arrived at the hotel, I asked he please give my thanks to Chief Inspector Ji. Constable Ming nodded, so I assumed he understood enough English to comply.

Terri was waiting for me when I arrived in the suite. I gave her a rundown on the day and how unsuccessful it had been. She did the same for me. Evidently, Ann had called her after her meeting with Mr. Jianguo and asked if Terri would like to go to police headquarters with her. Terri had agreed.

Ann told Terri her meeting with Mr. Jianguo was very unproductive. Sun believed the issue with Dennis and him not being a part of the murder hurt John. The fact that the eighth life test had not been located also made it appear John either killed Mr. Zhu or arranged for it. Sun was very pessimistic about John being released and was hoping Terri or I would find a breakthrough.

Terri said she waited in the lobby of the police station while Ann visited with John. The visit lasted less than thirty minutes, and Ann was very quiet after her meeting. All Ann would say as they rode in the taxi to the Grand was she missed her husband and felt totally helpless.

"Where is Ann now?" I asked.

"She said she had to meet a client tonight for dinner."

"I thought she might have dinner with us," I said. "I could, at least, make sure she knows everything we are doing to prove John is innocent and for her to not get further discouraged. We will find a way to prove John did not commit any murder."

Terri answered, "I believe she continues to immerse herself in work to keep from thinking about it."

"Are you hungry?" I asked.

"You know me; I can always eat," Terri said, laughing while going into the bedroom to change and put on makeup.

I was so frustrated with this whole thing I really didn't want to change for dinner, so I did not!

"Where to?" Terri asked.

"Chinese or Western?"

"Oh, come on! Please don't make me decide. If you do, we will go to Zhangshengji's," Terri said.

"Okay, okay. How about you!" I said.

"Only if you can catch me!" Terri said. She made a break for the living room, but a well-placed pillow throw hit her in the back of the knee, slowing her down enough for me to catch her.

"I give up, I give up," she said, laughing and running back into the bedroom, falling across the bed.

I followed, falling across her and the bed. "This is too easy!"

"I felt sorry for you being older and all!" she replied.

"That was cold!" I said. "I guess you will have to pay."

The payment was quick but creative, causing Terri to say afterward, "I take back what I said about you being old!"

We decided to try a new restaurant also close to the hotel. One Bin had recommended when we last saw him.

Xiao Zhou Xian was located outside the Shi Ji Ming Yuan apartment complex located next to the Guanghua Bridge. The apartment complex overlooks the river. The restaurant is family owned and has been in business for over thirty years. One of the current owner's sons was a university classmate of Bin. Most Westerners don't find Xiao Zhou Xian, and most people who work in the restaurant don't speak English.

We spent a few minutes trying to decipher the menu, getting nowhere. Our Chinese was not fluent enough to read the characters. Fortunately, an Englishman who had lived in Changzhou for many years and understood how to speak and write Chinese befriended us.

"May I be of assistance?" he asked. His accent was pure London. As Terri said later, "I could have listened to him all night long!"

"You appear to be somewhat lost in the menu."

"You noticed, huh?" I said, laughing. "Yes, we sure could use some help." He appeared to be in his sixties, tall, slim, with gray hair. His appearance and carriage portrayed a proper Englishman.

"Then may I recommend a dish for each of you before I leave."

"We would certainly appreciate it," I replied.

He walked to our table and sat down. "Before we begin, I am Reginald Waithwrite. I am attached to the UK Embassy here in Changzhou."

"It is a pleasure, Mr. Waithwrite," I replied. "My name is Rick Watson, and this is my wife, Terri."

"First trip to China?" he asked.

"No, we have been here several times for business and pleasure. A close friend recommended this restaurant."

"A Chinese friend?"

"Yes, he is acquainted with the current owner's son," I replied.

"Ah, I see. Well, let's pick something for each of you that symbolizes the restaurant," he said, looking over the extensive menu. "For you, Mrs. Watson, I recommend the Xijang lamb and chili grill. It has cubes of lamb marinated in lemon juice, garlic, and chili oil. You do like lamb, don't you?"

Terri said, "Yes, I do. The dish sounds wonderful."

"Excellent, now something for you," he said, turning his attention my way. "I hope you don't have an aversion to salt?" he asked.

"No, I don't," I replied.

"Excellent, then I suggest for you the spicy salt and pepper spare ribs. They are my personal favorite, not to slight the lamb, which I also find hard to resist. The ribs are prepared using China Sea salt and Szechaun peppercorns with the chef's special five-spice powder and then deep fried."

I enthusiastically nodded my agreement.

"For a vegetable, take the bok choy," he continued. "The bok choy is flavored with a red wine vinegar and two special soy sauces also developed by the chef."

Terri said, "Everything sounds just wonderful. We don't know how to thank you. You were right, we were struggling with the menu."

"No thanks needed," he replied. He called over one of the waiters and placed our order. "Now I must be on my way. I trust you will utterly enjoy your dining experience at Xiao Zhou Xian. Maybe we shall meet here again."

I thanked him profusely and said, "I certainly hope so. Maybe we can return the favor sometime."

"It would be my pleasure to allow you both to do so," he said. "Good evening and bon appétit." He tipped his hat and made his way to the exit.

"Now that was a true Englishman," Terri said as he walked out of the restaurant.

Our dinners were more that he advertized. We had once again found outstanding local food we would long remember.

It was after 11:00 when we arrived back at the hotel. It didn't take long for us to get ready for bed and pass out in our comfortable bed still full from dinner.

What the hell? . . . Terri's cell phone was ringing. I shot up and looked at the alarm clock beside the bed. It was 3:14 in the morning.

Terri picked up her cell and turned on her nightstand light. The cell phone continued to ring in an almost panic mode.

"Hello . . . Hello? What? Who is this? Calm down and speak slower, please. I can't understand you. Ann, is that you? What's happened? Is something wrong? Oh, my God! Are you sure?"

She turned to me and said, "It's Ann. She is saying Gregory has had an accident, and he's dead!"

"*What*? Put her on your speaker."

Terri laid her cell down and pushed the speaker button.

"Ann, honey, this is Rick. Can you tell us what happened?"

"Oh my god, Rick," Ann wailed. "This is horrible, just horrible!"

Ann continued to sob uncontrollably into the phone, but did manage to answer me. "He had taken off from the airport where he keeps his ultralight. He crashed across the road off the end of the runway onto a golf course. According to Kim Hancock, his HR person, he was dead when the first people got to the accident."

"Where are you?" Terri asked.

"At the hotel," she said, sobbing harder.

"The Shangri-La?" I asked.

"Yes."

"What room?" Terri asked.

"Room 1802," Ann replied.

I asked Terri, "Isn't Dennis back in Shanghai?"

"I believe so."

"Ann, I will call Dennis and have him come to your room," I said.

"Okay," Ann replied, her voice barely audible.

"Stay put and hang in there. I will make some calls and see what I can find out and call you back, okay?"

"Okay."

Terri said, "Honey, everything will be okay. Have you told John?"

"Noooo!"

"We will take care of that," I said. "I don't want him hearing this from Chief Inspector Ji."

"Stay put. We'll call you back," I said as Terri hung up her cell.

I placed a call to Dennis. He answered on the third ring. "Dennis, this is Rick. Are you in Shanghai?"

"Yes, I am," he replied. "What's up? It's the middle of the night here. Is something wrong?"

"Are you in the hotel?"

"Yes," he said. "What's going on, Rick?" His voice had the first sounds of panic.

"Terri and I just hung up from talking to Ann Alworth. It seems Gregory was killed this afternoon in an ultralight accident somewhere around Charlotte."

"*Do what!* Are you positive? That can't be possible. Gregory Brightson knows more about ultralights than almost anybody in the country. He has been flying them for at least fifteen years. Did this happen at Rowan?" Dennis asked.

"Right now I don't know. All Terri and I know is Ann called to tell us Kim Hancock called her within the hour to break the news. I assume you have not heard from Ms. Hancock?"

"Nor anyone else. Is Ann in the hotel?"

"Yes, and I would like for you to go to her room and see what you can do to calm her. She is really upset," I said.

"I am on my way. I should be there in about ten minutes. What's her room number?"

"Room 1802 on the executive level," I replied. "Look, Dennis, I don't have a clue why this happened. It may truly be an accident, and it probably is, but the timing sure sucks if it is. I am going to call Kim Hancock after I hang up with you and see if I can find out more. I will call you back either way after I talk to her. Try to keep Ann calm until I get back to both of you."

"I will stay there until I receive your call," Dennis said. "This seems surreal. Right on the precipice of his big event, too."

Next, I placed a call to Changzhou police headquarters. It was 5:00 a.m. "Ni hao, Station Sergeant Lu," a voice on the other end of the line answered in Chinese.

Before he could complete his first set of words, I asked for Chief Inspector Ji. The station sergeant replied, "He is on his way in here, sir."

"Thank you," I said. "I will try his cell."

"Ni hao, Rick. Why have you called me so early? There must be big problem," he said.

I began, "Chief Inspector Ji, I have really bad news. Gregory Brightson was killed in a flying accident early in the afternoon in US time—in China time, around three this morning. I need to talk to John Alworth."

"I am so sorry, Rick. I know this is a blow to your attempting to find John Alworth innocent. How did it happen?"

"All I know now is he was flying his ultralight from an airport near Charlotte some time prior to two in the afternoon, Charlotte time, and his plane crashed right across the road from the airport onto a golf course."

"Rick, I'm sorry to hear you say that. I will make sure John calls you immediately. Any idea to the cause?" he asked. "Any evidence of foul play?"

"Not at this time," I said. "I hope to have more information later this afternoon or early evening time here."

"I will be in my office in about ten minutes," he said. "Always heard those types of aircraft were death traps, anyway," he continued.

"They may be; however, I don't know much about them," I replied.

"I guess you will now," Chief Inspector Ji replied.

Good to his word, Chief Inspector Jihad John call within the ten-minute time frame.

"Rick, this is John. What's going on? Has something happened to Ann?" his voice was quivering.

"Ann is okay. She is safe in her room at the Shangri-La," I said. "It's Gregory. He was killed in an ultralight crash this afternoon outside of Charlotte."

"He was what? You're joking. Gregory dead! That's not possible. He has flown those things for years, and no one is more conscientious about safety than him," John said. His voice was disbelieving. "Who told you?"

"Ann called Terri on her cell at three fourteen this morning from Shanghai. Apparently a Ms. Hancock in HR from WMT called her," I said.

"Why would Hancock call Ann and, wait a minute, why would anyone call her to begin with?" John asked.

"Those are good questions, buddy, that need answers," I replied. "I am going to call Ms. Hancock right after I hang up. Maybe she can shed some light on this thing."

"You don't think Gregory's crash has anything to do with Zhu's murder, do you?" John asked.

"I really don't see how or why, but I do know the timing sure sucks! I hope to know more after talking to Ms. Hancock. I will get in touch with you as soon as I know more."

John asked, "How was Ann? She was very fond of Gregory."

"She was pretty broken up when she called," I replied. "I asked Dennis to spend some time with her to calm her down. I believe Terri is going to Shanghai this morning to relieve Dennis and stay near her for a few days."

"I know she will appreciate that. You know how emotional she can be. It's the Italian coming out in her!" John said.

"Hang tight, buddy; I will call you back as soon as I can with more."

"What options do I have?" John answered.

Terri had been on her cell making reservations to Shanghai and had booked a 7:00 a.m. train. She had called Ann to update her. She said Dennis was there with her and said he would stay until she got there.

"John had a few real good questions for me to ask Ms. Hancock when I talk to her," I said. "And the more I think about it, why was Ann so emotionally distraught over this?" Before Terri could answer, I said, "John attributes it to her Italian heritage."

"I was wondering that myself. I know she and John were close to Gregory since he really didn't have any relatives or many friends around him. Maybe that's why," Terri said.

"Well, we can't worry about that now," I replied. "For now, we need to gather as many facts about what happened as we can and decide what to do next," I said.

"My train leaves in an hour, so I had better get moving," Terri said. "I still have to pack. How long do you think I should plan on being in Shanghai?"

"I would think no more than a couple of days," I said.

"I'll plan on three to be on the safe side."

Terri headed for the bedroom to pack, and I placed a call to Ms. Hancock. It was 6:15 in the morning. She answered almost immediately. "Hello, this is Kim Hancock."

"Ms. Hancock, I don't believe you know me. This is Rick Watson, and I am calling from Changzhou in China regarding what has happened to Gregory Brightson."

"Oh yes, Mr. Watson, I actually do know who you are. Mr. Brightson has spoken highly of you of late and appreciated what you and your wife were doing for poor Mr. Alworth."

"Excellent. I guess this has been an extremely trying day for you. Do you mind taking a few minutes to answer a few questions?"

"No, not at all," she replied. "I will do all I can to assist in finding out what happened to Mr. Brightson. I have already received a call from our board chairman, Mr. Conroy, and he had many questions as well."

"Can you tell me what happened around the crash? What airport was he flying from and exactly when did the accident happen?"

"Everyone in the company knew Gregory was an avid ultralight pilot. He had been flying them for years and had even built a few. His ultralight was hangered at the Rowan County Airport near Salisbury. I believe he was currently flying the Quicksilver MX Sprint." Ms. Hancock continued, "Rowan County is a very small airport, and that's why Gregory liked it. He felt safe leaving his ultralight there because he knew everyone who worked there."

"Do you know what happened to cause the crash?" I asked.

"Jim Chatman, the police chief of Salisbury, told me as this point everything was pointing to a mechanical failure."

"Did he say where or what type?"

"No, he didn't. Mr. Watson, this is just so hard to believe. Mr. Brightson was everything to this company. We were his family, his creation, and we were right on the brink of fulfilling his dream. This can't be happening!" Ms. Hancock said.

"Do you have a number for Chief Chatman?" I asked.

"Yes, just a moment, I have it in my file. Here it is. His number is (704) 555—4000."

"May I go back to an earlier question? Why did the chief call you initially after he arrived at the scene of the accident?" I asked.

"You see, Mr. Brightson and I had an agreement. Since he had no real family he could count on, he used me as his emergency contact. He kept a card in his wallet with my data and company data on it with instructions for anyone who found it to call me. It is my job to keep everyone informed and connected," Ms. Hancock said.

"And why did you in turn call Ann Alworth?" I asked.

"I originally attempted to call Mr. Dureno, but he didn't answer, so I called Mrs. Alworth. I knew she was in China, and I wanted Mr. Alworth to know as soon as possible."

"That clears up some earlier confusion," I said. "We were having a hard time understanding why there were calls to you and Mrs. Alworth."

"I know Chief Chatman personally after working with him and his team on things like employee background checks and other employee issues that involve the Salisbury Police Department," she said.

"Where is Gregory now?" I asked.

"He is at the city morgue in Salisbury, according to Chief Chatman," she answered.

"Do you happen to know where the ultralight crashed?" I asked. "And where it is now?"

"I believe Chief Chatman said it crashed right off the end of the runway onto the thirteenth fairway of the Rolling Hills Golf Club," she answered. "It's ironic, isn't it? He died on fairway number thirteen, and he was always so lucky!"

I was not sure how to respond to what she said, so asked once again, "Do you know where the ultralight is now?"

"No, I don't, but I am sure Chief Chatman will know."

"Thank you, Ms. Hancock, for being so helpful and taking the time to talk with me. I know how you must feel being placed in a position like this," I said.

"Mr. Watson, it is my job to handle situations like this; however, this is the first time in my career I have had someone like Mr. Brightson die. I sure hope we find out why this terrible accident happened."

I thanked her again and started to hang up when she said, "Before you go, Mr. Watson, how is poor Mr. Alworth doing? No one here believes he could do something like what he is being accused of. Are you going to prove he did not so he can come home?"

"We are certainly trying, but losing Gregory is a blow to our strategy. I assume the board will appoint someone as CEO, at least temporarily?"

"According to Mr. Conroy, the board of directors will be meeting early next week to appoint someone," Ms. Hancock answered.

"Do you have any succession planning for an event like this, and if so, are you at liberty to tell me who was Gregory's backup on that plan?" I asked.

"Yes, Mr. Watson, we certainly do have a succession plan; however, I am not at liberty to divulge who that person is. I can say this: the board has asked Mr. Dureno to be at that meeting."

"Ms. Hancock, thank you again for your help. I will be back in touch in a couple of days as events unfold at WMT."

My next call was to Chief Chatman of the Salisbury Police Department.

"Salisbury police, Sergeant Acton speaking."

"Sergeant Acton, this is Rick Watson. Is Chief Chatman available?"

"Yes, sir, he is; may I ask why you are calling?"

"It is regarding the death of Gregory Brightson earlier this afternoon."

"Yes, sir, I will put you right through."

"Chief Chatman, may I help you?" Chief Chatman's voice was a deep official police voice that I learned during our conversation was honed over many years of police work.

"Chief Chatman, this is Rick Watson. I am calling from Changzhou, China. Do you have a few minutes to talk about the ultralight crash that took the life of Gregory Brightson this afternoon?"

"Yes, I sure do, Mr. Watson. May I ask why you would be calling me all the way from China?" Chief Chatman asked.

I spent the next fifteen minutes describing WMT, why I was in China, about John Alworth's role with the company, Dennis Dureno and his role, the new product and its production in China, the failed rogue life test, the murder of Mr. Zhu, John Alworth being charged with the murder, and the connection to Gregory Brightson.

"My, my, sure looks like you have your hands full, Mr. Watson."

"Let's just say it has been very frustrating trying to find the eighth life test and to prove Mr. Alworth is innocent," I answered. "Chief, is there any evidence to indicate the crash was anything more than either pilot error or mechanical failure?"

"No, sir, there isn't. My guys have been over the crash site and the ultralight with a fine-toothed comb. We can't find a thing wrong. I believe the poor bastard screwed up during takeoff and took a nose dive onto the golf course."

"The fact that you can't find anything physically wrong with the ultralight bothers me, Chief," I said. "Mr. Brightson was a very experienced, well-known and respected pilot. Everyone I have talked to today has confirmed he was an incredibly detailed person during any preflight inspection. They believe he would not have taken off with a potentially faulty ultralight."

"Mr. Watson, I have been involved in police work for over thirty years, and I have been chief of police in this town for fifteen of those thirty years. We may be a small police force, but we have the latest technology and a well-trained force capable of working this type of case."

"I apologize if I offended you or questioned the integrity of your force," I said. "What has happened is very strange under the circumstances. Chief, would you allow me to have one of my friends spend some time with you and your guys on this one?"

"Mr. Watson, I hate to be stubborn about this, but I really don't think I need any help from some big-city investigator to assist us in completing this case," he said. "Who did you have in mind?"

"I have a friend who is also a police chief in South Carolina. He and I were in the marines together and solved a pretty big drug case," I said. "His name is Gary Hildegard."

"What did you say his name was?" Chief Chatman said. His voice was filled with excitement.

I answered, "His name is Gary Hildegard. He is the retiring police chief in Beaufort."

"Well, I'll be dammed! He is welcome here anytime," the chief answered. "We have been police friends for years. He is one smart cookie. Hell yes, bring him on."

"That's fantastic, Chief," I said. What a break! Surely we would make some headway on finding out what really happened to cause the crash. "I plan on calling Gary right after I complete our call."

It was 8:30 at night when I placed my call to Gary Hildegard.

"Hello."

"Gary, this is Rick Watson."

"Rick, well I'll be damned. I haven't talked to you in a while, other than that last e-mail. How the hell are you? Did you ever solve that case you were working on? Probably not, or you wouldn't be calling me on a Thursday night at eight thirty."

"First of all, I am calling you from China. It's eight thirty on Friday morning here. Second, as you know, Terri and I have been here for about a month trying to help our friend John Alworth. Appreciate your earlier help. Continuing story here. Until today, I wasn't sure if we were going to prove he is innocent; however, something has happened in the United States that may begin to unravel this case."

"Yes, I know John," Gary said. "Remember, we met him at one of your July 4 parties at the lake. He worked for that startup company in Charlotte, didn't he?"

"That's him," I said. "The company is WMT, and he is the director of quality. His boss, Gregory Brightson, was killed in an ultralight crash in Charlotte this afternoon. I need your help to investigate what did happen. So far, the police are calling it an accident, but I am not so sure."

"Where did this happen, and who is doing the investigation now?" Gary asked.

"He crashed his ultralight on a golf course off the end of the runway. He took off from Rowan County airport outside of Salisbury," I said.

"Salisbury, that's Chief Jim Chatman. Chatman's a good man. I have been involved with him in several national committees. Does he know you are talking to me?"

"Yes, he does," I answered. "In fact, he is looking forward to your support."

"Good. I am three weeks from retirement, my replacement is in place, and I have some time off due me. I would be glad to help. I assume yesterday!"

I laughed, "Yes, right now. That's one thing I have always admired about you, never one to run away from a fight."

"Bring me up to date," he said.

I spent the next twenty minutes describing everything from our arrival in China to Gregory's crash. I spent time detailing the new product being produced at UCPC, the lost eight life test and the possible consequences of that test, what would happen to the company and its people if the product

265

did not begin production, who gained and who potentially lost, and why John Alworth felt the crash might not have been accidental.

"I believe I have enough to get started but will probably need a memory refresher as we go along," Gary said. "Why don't you come back to Charlotte and do this investigation yourself?"

"I would, but I have requested a ninth water sample from WMT so I can run an additional life test to either prove or disprove there is a problem, and I am trying to track down sample number eight. If I can find the data on that test, it may lead to the real killer. I can't think of anyone better than you to help me if I can't be there myself."

"It will be fun working together again. I was wondering what happened with the information on that guy Dureno. I'll clear the decks tomorrow and be there sometime Saturday morning. Should I call Jim and let him know I am coming?" Gary asked.

"Please do, and Gary, you will get anything you need on this one. Dureno is not our guy, even though I thought so for a while. Figuring out what really happened is critical for the company and for me to get John out of China a free man. I really appreciate you taking time to help."

"Anything for you, my friend. I will be in touch after I get settled in Salisbury."

We disconnected. Terri had already left for the train station, placing a note under my nose saying goodbye and giving me a kiss while I was on the telephone. The note said she would call me from Shanghai.

After 9:00, no food, no coffee, my next stop was the executive floor lounge prior to them closing. After breakfast, I planned to call Mr. Da and arrange a meeting to discuss the day's events and obtain approval for the ninth life test.

Ms. Xu said Mr. Da was in the plant and would be happy to meet with me at 1:00 in the afternoon. She said the UCPC van would pick me up at 12:30 in front of the hotel.

By noon on Friday, May 30, it was already hot and smoggy in downtown Changzhou. The driver was waiting, and we headed for the plant.

Ms. Xu greeted me as I exited the van in the courtyard of UCPC. She said Mr. Da was waiting for us in the conference room.

"Ni hao, Mr. Da," I said, bowing slightly as I shook his hand.

"Ni hao, Mr. Watson," he said, returning my bow.

"Mr. Watson, Mr. Da received call from Mr. Dennis Dureno this morning telling of Mr. Gregory's death. Mr. Da very sad to hear news," Ms.

Xu said. I looked over at Mr. Da, and he nodded to me with a very sad look on his face.

"Very bad," Mr. Da said.

"I am sure Mr. Dureno discussed with you what the plans were going forward?" I asked Ms. Xu.

"Yes, he say WMT stay on present course until further notice."

"Thank you, Ms. Xu, for the information. Today, I would like to obtain Mr. Da's agreement for me to supervise, along with Billy, another life test."

Ms. Xu translated, "*Li cha Xinag Sheng Qi Wang Nin Pi Zhou Zai Duo Yi Ci De Shu Ming Shi Yan.*"

Mr. Da nodded and said, "Shi, shi."

"Please tell Mr. Da I will begin the life test today if the water sample has arrived."

"Water sample has arrived in plant," Ms. Xu said.

She translated my request to Mr. Da, and he agreed.

I promised to give them daily updates on our progress. I thanked them and excused myself to go to the plant and locate the water sample and Billy.

Billy was working in the QC lab when I found him. I described what I wanted to do with the new water control sample. He agreed with my premise and called someone to retrieve the water sample from receiving department.

We were walking toward the initial test unit where all the previous life tests had been conducted when my cell rang. It was Terri. I excused myself and walked back to the QC lab where I had better reception.

"Okay, honey, I am in the QC lab," I said. "I assume you made it to Shanghai."

"I did," she answered. "The train was on time, and I arrived at the hotel around nine fifteen, just in time to say goodbye to Dennis. He said there was going to be a board of directors meeting on Tuesday, and they requested he be present."

"I know. Ms. Hancock told me he had been invited to attend. How is Ann?"

"She's okay. Probably a little mad at this minute because she thinks Gregory's death will not help John."

"She's right, but time will tell. Has Ann heard when Gregory's funeral will be?" I asked.

"No one has called or e-mailed anything as of now," Terri said.

"I have Ms. Hancock's cell number. I will call her late tonight and see if arrangements have been made and for what day and call you back," I said.

"My guess is the funeral will be on Monday, June 2, in Charlotte," Terri said. "If that is when the funeral is, Ann and I will probably fly to Charlotte on Saturday late morning. Are you planning on attending with us?"

"I don't think so. Billy and I will have the ninth life test up and running by then, and I don't want to miss any part of the test. Both of you being there will be sufficient, I believe."

Terri said, "Ann is calling Sun this afternoon to see if the Changzhou police will release John under the control of Changzhou, Huaide, and Zhanglinfang and Mr. Xue, their president, so he can attend the funeral. If that happens, would you change your mind?"

"No, and in fact, I would rather John was here to direct this life test," I said.

"Honey, I am going to e-mail you a list of clothes I will need for the trip. Will you please have them delivered tomorrow early in the morning by courier?"

"Okay, I should be back in the suite early enough to have them sent out in the late afternoon run, and you should have them no later than eight. When do you think you will be flying back to Shanghai?" I asked.

"Ann wants to be in Charlotte to hear the results of the board of directors meeting, so I guess Tuesday afternoon. That should put us back in Shanghai Wednesday afternoon."

"Honey, I miss you already, and you are still in China," I said.

"I miss you, too, babe."

"I'll call you after I have talked to Ms. Hancock," I said.

"Bye, honey." Terri hung up.

Billy and I spent the balance of the day completing about 75 percent of the setup to begin the life test on the initial production machine.

We agreed to begin at 8:00 the next morning; however, Billy could only work a half day on Saturday because he had commitments in the afternoon and evening in Shanghai.

I arrived back at the Grand at 5:30 that afternoon. I checked Terri's e-mail and packed everything on her list, plus a few things I thought she might need that were not on the list. The courier picked up the suitcase at 6:30, promising the delivery would arrive at the Shangri-La before 8:00Saturday morning.

Terri did not answer my call, so I left her the confirmation number and name of the delivery company, promising to call her again after I had talked with Ms. Hancock.

I had a lot on my mind and decided to have dinner sent to the room. There was much going through my head; it was becoming impossible to keep all my thoughts straight. It was time to develop a list. I called my lists my BOM (bill of material), a term taken from my days in manufacturing. Every product made has a bill of material for all the components, subassemblies, and assemblies, so I naturally gravitate to making lists. Terri hated my lists, preferring to play the memory game. She wasn't always successful thinking of everything we needed. When that happened, I put my list in her face and ran for my life!

Ten o'clock Friday night arrived while I was deep in thought. I had set my alarm clock to remind me to make my call. Ms. Hancock was in her office.

"Good morning, this is Kim."

"Good morning, Ms. Hancock. This is Rick Watson."

"Good evening, Mr. Watson."

"Ms. Hancock, have you been informed what day plans have been made for Mr. Gregory's funeral?" I asked.

"Yes, I have, Mr. Watson," she said. "You see, I am making the arrangements. The funeral will be at ten a.m. on Monday morning at the Covenant Presbyterian Church located at 1000 East Moorehead Street in Charlotte. Are you planning on attending?"

"No, I must remain in China; however, my wife Terri and Ann Alworth will be in attendance."

"Have you learned any more about Gregory's accident, Mr. Watson?"

"No real developments as of now, Ms. Hancock; however, my good friend Gary Hildegard, police chief in Beaufort, South Carolina, will be arriving on Saturday morning to work with Chief Chatman."

"That's encouraging," she said. "I am sorry I will not get to meet you, but I look forward to meeting Mrs. Watson."

"I am sorry I won't be there as well, but we have another controlled life test started, and I need to be here for that. I won't keep you, Ms. Hancock. Thank you for the information on the funeral."

"Goodbye, Mr. Watson, I'm glad I could be of assistance."

I hung up the phone and called Terri on her cell. It was after 11:00, but I knew they would not be sleeping. This time she answered.

"Hi, honey. I received your voice mail. Ann and I were at dinner, and I did not hear the phone ring. Have you talked to Ms. Hancock?"

"I just hung up from speaking to her. The funeral will be held on Monday at ten a.m. at the Covenant Presbyterian Church. I will send you an e-mail as soon as we hang up with the address. Ms. Hancock is in charge of the arrangements and will be expecting you."

"Thanks, honey. Ann talked to Sun this afternoon about having John released so he could attend the funeral. The judge refused the request."

"That's too bad, but not unexpected," I said. "I know Chief Inspector Ji would not support it if he were asked."

"Apparently, he didn't."

"Your flight is at two tomorrow, is that right?" I asked.

"Yes, it is," Terri answered. "We have a direct flight on US Air and arrive in Charlotte at one thirty in the afternoon on Saturday. You know what you always say, 'If you don't like the day you are having, fly to the United States from China and relive a portion of it'!"

"I bushed," I said. "I am going to the plant early tomorrow morning to work with Billy. Call me before you take off."

"I will, babe. Love you!"

"Love you, too. Goodnight."

"Goodnight."

My head hit the pillow, and I was asleep. I awoke to my alarm clock telling me it was time to begin my morning routine before heading to the plant.

Chapter 21

Rain was pelting down when I exited the hotel to the waiting UCPC van on Saturday morning, May 31. Today's rain was probably the hardest rain I had ever experienced in China. The driver was struggling to keep the van on his side of the road with the torrent of rain cascading across all traffic lanes. He took his time and let out a sigh of relief when we entered the courtyard at UCPC. He pulled the van as close to the entrance of the first production building as he could.

I made a mad dash for the door, watching to assure I did not slip on the marble floor as I entered the building. I walked through two other production buildings before reaching the QC lab. Billy was already in the lab working on entering data from control water sample number nine.

"Good morning, Mr. Rick. You manage keep dry?" he asked.

"Good morning, Billy," I answered. "Thank goodness the first production building's door is close to the main driveway. Your driver took pity on me!"

Billy laughed and said, "You are guest here. He must do that."

Billy and I spent most of the morning preparing the process plan and entering data in order to begin the actual life test in the afternoon. We selected the eighth production system because it was the one supposedly used for the unapproved eighth life test. Our goal was to complete the original life test plan and then continue on testing until we were comfortable we had proven there was or was not a fatal flaw in the system. UCPC had produced ten preproduction units to date.

We completed everything by noon and decided to push the button to begin the life test on Monday morning. Billy and I left the QC lab together. The UCPC driver was waiting for us right outside the same door I had entered earlier that morning. The rain had not let up as we entered the van.

The driver would take me back to the hotel and Billy to the train station to catch his train to Shanghai.

I said goodbye to Billy when the driver dropped me off at the Grand. I thought about Terri as I was entering the suite. Her flight would be on the runway beginning its taxi for the flight to the United States if everything was on time.

Lunch was next on my list, and then maybe a sauna. I felt tired. Maybe being away from Terri and knowing that she would be close to Lake Marion made me a bit homesick.

The sauna did its job, and I was completely relaxed when I finished. A quick dip in the pool brought me quickly back to life. I thought about setting up a meeting with or calling John, but I really had nothing new to tell him. It was a good afternoon to catch up on my reading and my favorite Internet sites with the rain pounding on the suite windows. Somewhere along the way, my eyelids gave up, and I took an unplanned nap.

Most of my life, I believed a nap during the day was a sign of getting old, and I certainly fought that concept. I had learned over the past few years that our very active physical routine allowed some downtime, so I occasionally gave in. Terri was kind not making an issue of it.

When you are used to sharing dinner with someone you love, eating alone is not fun, plus it was still pouring outside, so I chose the Western restaurant in the hotel. It was convenient, and the food was acceptable. The telephone in the room was ringing when I entered the suite at 10:00 that evening.

"Hello," I said.

"Rick, this is Gary."

"Hey, Gary, how's it going there?"

"I just got here and have had my initial briefing with Jim Chatman and his team. This afternoon, I should be able to visit the crash site and go to the airport. I will call you back before the day is over and before you go to bed."

"Sounds good to me," I said. "Happy hunting. I'm counting on you to help me get to the bottom of this."

"We'll get there. Give me a few days. Call you later."

It didn't take long to fall asleep with the rain hitting the window. The ring of my cell phone woke me from a deep sleep. I looked at the alarm clock: it was 2:30 Sunday morning. I knew immediately who it was.

"Good afternoon, honey," I said, answering the phone. "I assume you both made it."

"Sorry to wake you in the middle of the night, but I knew you would be worried if you woke up in the morning without hearing from me," Terri said. "Yes, we are here. The flight was smooth and on time."

"Good, what are your plans for the rest of the day and Sunday?"

"We are both tired, so I guess some sleep. Then find out what we can about Monday. Ann is going to visit some of her clients on Sunday, and I thought I would call my aunt Susie and spend some time with her."

"Give her my best, if you do get to see her," I said. "I miss you, honey. It doesn't seem right for me to be in China and you in the United States."

"I miss you, too. You're right; it doesn't seem right. Let's hope we can bring this to a head soon and both get back home. I love you. Go back to sleep. I will call you tomorrow."

"I love you, too, goodnight or good afternoon!"

For the second time, Sunday morning, June 1, arrived late in the suite because of Terri's call earlier call. I headed for the gym to exercise and get in a good long run. The rain had stopped, but the traffic would prohibit a safe run outside. I picked up some fruit and Danish in the executive lounge prior to returning to my room. My cell rang at 10:00. That would be Gary. He promised to call me at the end of his day on Saturday.

"Morning, Rick."

"Good evening, Gary. How has the day gone?" I asked.

"Let me give you a recap and where I think we are with this thing," he answered. "I arrived late this morning and talked to Jim Chatman and read his reports to date. Then I visited the crash site with one of his detectives, and then we went on to the airport to view the ultralight. There are three other ultralight owners who fly out of here. They were there looking over Gregory's ultralight when we arrived. I spent about an hour with them discussing why this could have happened."

"Did they have any ideas?" I asked.

"Oh yeah, they took me through the entire process of preflight. They told me when a pilot does a preflight, it begins at one location and systematically works around the entire ultralight, back to the original starting position. Everything is more out in the open than, say, a Cessna. You can see all the cables, fuel lines, and everything else. They all agreed Gregory was a very thorough and conservative pilot. Gregory flew a Quicksilver MX Sprint. Please bear with me, because I am going to read the rest from my notes to make sure I get this right. The Quicksilver Gregory flew was a single seater. It weighs 254 pounds with a maximum fuel capacity of five gallons. His Quicksilver could fly at a maximum of fifty-five knots in level flight, with a stall speed of twenty-four knots. Are you still with me?" Gary asked.

"Yes, I understand," I replied.

"Well, there is more," Gary continued. "Ultralights are considered vehicles by the FAA or NTSB, so many times, crashes aren't really investigated as they should be. That's the case here. The local police and I are the only ones interested in the crash.

"Now, here's the good part," Gary continued. "I am still using my notes; Gregory's Quicksilver used a Rotax 447, forty-horsepower, two-cycle engine and a wooden propeller. The pusher engine has a side mount exhaust on the right side of the engine, aiming straight back into the propeller spinning only a few inches away. The exhaust outlet is approximately one inch in diameter. The propeller would be about sixty-six inches in diameter, and full power, which Gregory would have had because of take off, would be turning 2,500 rpm. That causes a tremendous amount of centrifugal force, meaning somewhere around 3,500 pounds of outward pull on each blade.

"Still with me?" Gary asked.

"I'm not sure, but I know you know what you are talking about," I replied.

"Let's just say these guys have educated me enough to understand what they believed happened."

I asked, "What do they think happened?"

"They have reached a joint conclusion that something catastrophic happened within the exhaust system, causing an explosion that destroyed the propeller. The propeller pieces sheared one or maybe both of the tail tubes, causing the failure. When the propeller disintegrates, the lack of balance caused by centrifugal force would rip the engine off the airframe and cause the airframe to fail. The ultralight would come spinning straight down. There was nothing Gregory could do. He was too high to land, and besides that, he would have lost all control."

"Where are you? Do you believe their analysis?" I asked.

"I know Jim does, but I don't feel right yet. Everything they say makes sense, but I know you want to make sure no one caused this to happen, right?"

"Right on, my friend," I answered. "We have been there together before when we have had this gut feeling something isn't right, doesn't add up."

"I think I will call Larry Henry," Gary said. "He is an old friend in Beaufort that flies these things. Maybe he has some other ideas. Specifically, I'll ask him what he would do to cause a crash."

"I would appreciate that," I said. "If he reaches the same conclusion as the three guys there did, then I'll back off."

"I will call you on Monday. Tomorrow, Lance, my grandson, is in a baseball tournament in Charleston. The whole family is meeting me there tonight."

"That's terrific; wish him luck from me."

"I will. Call you Monday."

It took a few minutes to unwind from Gary's call and all the technical information he described. It was good to be working with him again. It felt like old times! Gary Hildegard was one of the best investigators I had ever known. His eye for detail that most people missed was one in a million, and his gut was calibrated like no one else I knew.

My phone sounded a text message had arrived. It was Terri, checking in before she went to bed. We exchanged a few messages, and I told her I loved her and goodnight.

I spent the rest of the day and early evening shopping and visiting several tourist sites in the city—something I would not ordinarily do if Terri were here. I couldn't resist going to Zhangshengji's for dinner. The owner asked me why I was alone and where his favorite customer was. I told him Terri was in the United States but would be back soon. Dinner was excellent as usual.

Chapter 22

Monday June 2 was a sunny day. The rain though intense had not lasted long; however, the rain left the city steaming in the morning. I was at the gym before seven and was able to complete a seven-mile run on the roads in the city. It was getting pretty crowed when I finished at just over one hour later. A quick steam and breakfast in the executive lounge finished off my morning.

The UCPC van was waiting for me at 9:00. The drive to the plant was uneventful—not like the ride in the rain on Saturday morning.

Billy was waiting for me in the QC lab.

"Morning, Mr. Rick. Better day today," he said.

"Morning, Billy. Yes, it sure is. Steamy, though," I replied.

Billy nodded and returned to his work.

"Are we ready to push the button?" I asked.

"Ready, Mr. Rick."

We walked to the production system number eight, and Billy pushed the start button. Our job for the rest of the day was to record data from the various checkpoints within the system and to measure the data from each checkpoint to the standard and the other seven life tests. The system performed to specification at each checkpoint during the day.

We completed the first day's testing at 6:00 in the evening. We performed the data run from that day and planned day two prior to leaving. The faithful UCPC van was waiting for me when I arrived in the courtyard to take me back to the Grand.

I settled for a sauna prior to dinner. It was a pain to do all these daily events without Terri. I really wasn't even hungry but did manage to grab a sandwich from the executive lounge before they removed the food in favor of chips, nuts, and drinks.

It was 10:00Monday evening before I noticed it. I was into my latest mystery novel when I thought about Terri at Gregory's funeral. I felt bad I could not be there but knew my place was here, trying to help John.

My cell woke me at 3:00 in the morning on Tuesday. *This had better be good*, I thought to myself.

"Rick, I hate to do this to you, but things are popping here!"

Gary's voice made me sit up in the bed and shake the cobwebs quickly. "What's happening, Gary? You must have found out something important," I said.

"More than important, buddy!" Gary said with excitement in his voice. "We were right about the crash. It was deliberate, and I have the evidence to prove it! How's that, big boy!"

"Whoa, slow down," I said. "You mean we were right the crash wasn't an accident?"

"No, it was not. I called Larry Henry last night, and he told me there was a way to rig an ultralight to crash that looks like a catastrophic failure. Let me read to you again from my notes.

"Larry said the Quicksilver has two one-and-one-quarter-inch diameter aluminum tubes attached to the trailing edge of the wing extending back to the tail. These are the main tail support tubes, and the propeller only clears them about an inch at each propeller tip. Larry said if someone wanted to create a crash, all they had to do was slide a three-inch-long piece of pipe of a slightly smaller diameter than the hole diameter of the exhaust outlet into the exhaust. It would be undetectable on any preflight inspection. He said a small amount of high-temperature silicone would ensure the piece of pipe would stay in place until well into the takeoff. If everything went to plan, the pipe would slide out at just the right time and fall into the propeller and cause the consequences as we discussed yesterday. Larry told me to look around the crash site for the piece of pipe. I did, and we found the pipe in the ditch on the airport side of the road, right where Gregory would have crossed over."

"My God, we were right!" I almost screamed into the telephone. "Someone did rig the ultralight to kill Gregory."

"And it gets better, my friend," Gary said. You know good things happen to good people? Well, a farm couple, Mr. and Mrs. Hackett, were on the way into Mooresville to do some shopping. They were travelling down Rowan Mill Road when Gregory flew over them. They saw the plane disintegrate and crash on the thirteenth fairway of the golf course.

"Mrs. Hackett saw a man with a pair of binoculars watching Gregory's ultralight. He got into his car and left as soon as Gregory crashed. Mrs. Hackett said the car was a new white Nissan Altima. She knew that because she and her husband have been looking for a new car and like that particular model. Mr. Hackett was able to get over half of the license plate number. Jim's boys are checking it out now."

"You have got to be kidding!" I said. "Son of a bitch, how did we get so lucky?"

"We are good people, my friend! Look, I have to go. I will call you later with more."

"I don't know how we can ever thank you, Gary," I said.

"Thank me when we get the bad guy," he answered.

"Okay, talk to you soon."

Chapter 23

Sleep eluded me after Gary's telephone call. I don't remember my last look at the alarm clock on the bedside table, but I woke to my cell ringing at 9:30 Tuesday morning, June 3.

"Hello, this is Rick Watson."

"Hello, Mr. Rick, I am Ms. Xu. Are you okay?" she asked. "Billy was worried you were not here yet and the driver said you did not come out of the hotel."

"Hello, Ms. Xu, yes, I'm okay. I was on the telephone to the United States very early this morning."

"Oh, so sorry for call then," Ms. Xu said. "I apologize for inconvenience."

"No need to apologize," I said. "Please tell Billy to begin and I will be there sometime after lunch and thank your driver for waiting for me."

"Yes, Mr. Rick. I so sorry wake you up, and I will tell Billy. Do you need driver later?"

"No," I answered. "I have another appointment and will take a taxi to the plant. Thank you, Ms. Xu. Goodbye."

"Goodbye, Mr. Rick."

I needed a hot shower to get my nervous system operating, a quick breakfast, and then to place a call to Chief Inspector Ji. I wanted to set a time to visit him this morning to inform him of the latest developments. I also wanted to see John to also inform him.

Chief Inspector Ji was in his office and agreed to see me at 10:30 as well as to give me an opportunity to see John right after our meeting. That left me just enough time to dress and find a taxi.

"Good morning, Mr. Watson," Chief Inspector Ji said as I entered his office. "I hope you bring good news."

279

"I do, Chief Inspector, and I hope the news is good enough for you to consider releasing Mr. Alworth on bail," I said.

"We shall see, Mr. Watson."

"A good friend of mine who is a police chief in the United States has been working with the local police where Mr. Gregory was killed, and he has discovered the crash of Mr. Brightson's ultralight was no accident. The crash was planned by someone who created a situation where the crash was inevitable. My friend and the local police, with the help of a local couple at the accident scene, may have identified someone who may be the killer. I hope to know that person's name and why he may have tampered with Mr. Brightson's ultralight by late tonight or tomorrow."

"Very interesting, but what does Mr. Brightson's death have to do with Mr. Alworth killing Mr. Zhu?" Chief Inspector Ji asked.

"I believe there is no way Mr. Alworth would have the opportunity to order someone to kill Mr. Brightson while he was in the United States. John is under constant surveillance here in your jail," I said.

"That may be true; however, I still fail to see the connection between Mr. Brightson's death and Mr. Zhu's death."

"You are not trying hard enough, Chief Inspector!" I said. "I have said all along Mr. Alworth had nothing to do with the murder of Mr. Zhu. That someone else who did not want the extended eighth life test to become public was responsible. When Mr. Brightson decided to stop production of the new product until the unapproved eighth life test was proven to be either correct or false, someone who had too much to lose made sure Mr. Brightson would not be in the way. John Alworth is not that person!"

"Suppose you are correct, Mr. Watson; who do you believe is responsible?" Chief Inspector Ji asked.

"At this moment, I am not sure, but when my friend and the Salisbury police locate the person the couple saw at the crash site, we will find out, and then you will be forced to release my friend for good."

"For your friend's sake, I hope you are correct, but for now, he must stay where he is. Would you like to see him now?" he asked.

"Yes, I would," I said. I was not happy with Chief Inspector Ji's inability to see things as I saw them. I had hoped I would be able to convince him of a reasonable doubt as to John's part in any killing. I would hit him again after I received word from Gary who the killer was and who may be ultimately responsible.

John was waiting for me with his customary guard when I arrived at the interview room on the fourth floor.

"Well, Mr. Watson, what brings you here? Have nothing else better to do than to harass your old buddy?" John said, smiling.

"Kiss my ass. If you don't want to hear what I have to say, I can be on my way!" I replied, also smiling.

"You're here now, so you might as well give me the news."

I spent the next forty-five minutes bringing John up to date on all the events of the past three days, and that Billy and I had begun the extended life test on the ninth sample. John was very quiet while I was talking and asked few questions. When I finished, he took a long breath, stood up, and paced the room for a few minutes.

"Do we have any idea who would kill Gregory?" he asked.

"Initially, I put my money on Dennis, but now, I am not so sure," I answered. "I know he needs the product to start production as soon as possible and the company to go public so he can cash in on his stock options. I believe he is in an untenable financial position at home, but he is too obvious. Besides that, the board of directors has asked him to be in Charlotte for the board meeting on Tuesday morning."

"Do you think the board is considering him as Gregory's replacement?" John asked.

"I believe so. I can't see any other reason for the board to specifically request he be there."

John said, "Maybe we have been looking at this the wrong way. Suppose there is someone who works for UCPC who would want to keep the product from going to production?" John asked. "Or maybe some other company has stolen the new product concept and paid someone inside UCPC to sidetrack or just plain kill production until the competing company can launch their competition product?" John asked.

"The only people in China who have contact with customers on a normal basis are Ms. Xu, Mr. Zhu, Billy, and Mr. Da, and I have seen no evidence of anything like that happening. I'm not sure about Charlotte," I answered.

"I assume you know Ann worked with Sun of Changzhou, Huaide, and Zhanglinfang in an attempt to obtain your release on bail so you could attend Gregory's funeral, but the judge said no," I said.

"Yes, Ann called me before her flight back to Charlotte," John replied. "She was pissed, and frankly, so am I. What's it going to take to convince these assholes I had nothing to do with what happened to Zhu?"

"If Gary and the Salisbury police can find the man the Hacketts saw, maybe he can lead us to the person responsible," I replied. "But, that would probably rule out anyone from here."

John sighed and said, "I guess nothing is going to happen for a while, but I really am tired of being here!"

"I know, buddy, but with Gary's help, we will get to the bottom of this soon," I said. "I hate to leave, but I must get to UCPC and continue the life test. I will either call or come by as soon as I know more."

"Let me know if you see any anomalies in the test; I may be able to help you interpret them."

"Will do. See you later."

John waved goodbye as the guard led him out of the interview room, back to his cell.

A taxi was waiting outside Police headquarters, and I arrived at UCPC at 2:00 in the afternoon. Billy was set up at the production system recording data when I arrived in the plant.

"How is it going, Billy?" I asked. Billy jumped as if he had just been shot out of a cannon.

"You scared me, Mr. Rick. I did not hear you come. I was into data reading and recording."

"Sorry, Billy," I said. "Are the numbers still within spec?" I asked. We had reached the halfway point of the normal life test time.

"Yes, Mr. Rick, all numbers are good. No problems."

Billy and I spent the balance of the day continuing to record life test data and comparing it to the standard life test data. Everything was performing as expected. We decided to call it quits at 5:00 in the evening and to begin the test again on Wednesday at 9:00 in the morning. We both walked to the courtyard where the UCPC van was waiting for me. Billy said he had additional work to complete in the main office.

I was about to close the door of the van when Billy asked, "Mr. Rick, will you be at hotel tonight?"

"Yes, I will, Billy. Why do you ask?"

"Oh, nothing, Mr. Rick," Billy answered over his shoulder as he was walking toward the main office building.

I thought the question was rather odd but quickly forgot about it as the van entered city traffic, and my mind wandered to Gregory's funeral. The ride back to the Grand was smooth as usual. I thanked the driver and agreed to be ready tomorrow morning at 8:30 to return to the plant.

A sauna was in order when I reached the suite, so I made a quick change and headed for the gym. A casual swim cooled me off, and a shower finished my early evening just in time for dinner. I missed Terri and really wasn't hungry. The Western restaurant in the hotel would suffice. The hotel's version of an American hamburger was my menu selection. The chef must have trained in or was from Belgium, because the burger was topped with a fried egg! The only place in the world I had seen that was Belgium. I said quietly to myself, "I can't wait to get home and have dinner at the J & E Bait, Tackle, and Home Cooking Restaurant about a mile from our house."

I was back in the suite by 8:30. I expected Terri to call me after Gregory's funeral at 10:00 in the morning Charlotte time, and maybe Gary would also have more to tell me before I went to bed.

The suite doorbell rang and surprised me. It was after 9:00. I had not ordered anything from the hotel and had no laundry to send out. I walked from the bedroom to the front door and opened it. Billy was standing on the other side with his head bowed, as if he was in trouble and afraid what might happen to him when I saw him.

"Billy, what are you doing here?" I asked.

Billy raised his head and looked directly into my eyes. "Mr. Rick, I must talk to you," he said.

"Please, come in, Billy," I said, opening the door so he could enter. "What is it I can do for you? Please, sit down," I said, pointing to the couch in the living room of the suite.

Billy quickly crossed the room and sat down on the edge of the couch. "Mr. Rick, I have something to tell you. It important."

"Billy, you can say anything you want to me, and I will do my best to understand and help you."

Billy sat there for at least two minutes. It was obvious it had taken a huge amount of courage for him to have come this far, but he wasn't sure if he was capable of going the rest of the way.

He finally looked up and said, "Mr. Rick, I have original and only copy of eighth life test."

"You what!" I said, trying to keep my voice as level as possible. The last thing I wanted to do was scare him off.

Billy began to cry. "I came into lab and saw Mr. Zhu lying on floor. He did not move when I call him. I knew Mr. Zhu run extended life test and it fail. He would not let me help him, but I saw report early that day. I was afraid if someone saw test, company would suffer and I lose my job. I took test and hid it. If no one find test, everything be okay."

I was stunned! It took a minute for me to recover to ask my question. "Billy, did you see who killed Mr. Zhu?"

"No, he dead when I saw him."

"Where is the test now?" I asked.

"I take test home to Shanghai."

"Is that where it is now?"

"Yes, Mr. Rick, I so sorry. I not mean cause so much trouble for you and Mr. John. Now Mr. Gregory dead, too," he said. "All my fault. None of this happen if I not take test."

I didn't know whether to kiss him or knock his head off. "Billy," I said. "Whatever happened is not your fault. Mr. Zhu was killed because someone did not want the data on the extended life test to be made public. If the test were made public, production would have been stopped until corrections were made. Based on how critical the results were, WMT might have missed the window to go public, and they might have gone bankrupt."

Billy continued to cry. "I not understand what you just say about WMT, but I want to give you test report."

"You say it is in Shanghai?" I asked.

"It safe in my apartment," Billy replied. He blew his nose, wiped it on his sleeve, and stopped crying.

"When can I get it?" I asked.

"I am very afraid to give to you in Changzhou or plant. It must be someplace outside where many people are."

"Why are you so afraid?" I asked.

"Mr. Da is violent man sometimes. I afraid what he may do to me if he know where you got test report. You promise to keep secret where report come from?"

"Yes, Billy. You can count on me. When and where would you like to meet me in Shanghai?"

"I will meet you tomorrow night, nine o'clock in the evening, in People's Square in front of theater. You come by yourself."

"Billy, I can't promise I will be by myself, but I will not let anything happen to you," I said.

"I trust you, Mr. Rick. I am so sorry I cause so much trouble."

I told Billy once again he was not the reason for what happened, and now, by giving me the test report, he would be part of the solution. That seemed to make him feel better, because a small smile crossed his face.

Bill thanked me once again and agreed he would not be in the plant on Wednesday morning. He was returning to Shanghai on one of the morning trains. I told Billy not to worry about the test we were currently running. It probably would not be needed now.

He shook my hand as he left the suite. "Thank you, Mr. Rick, you are good friend."

"Thank you, Billy," I replied. "I will see you tomorrow night."

I closed the door and stood there, not quite knowing what to do next! So, Billy had the data all along. The little shit was more worried about his job than anything else. I wondered what Chief Inspector Ji would say about these developments? If Billy was so concerned about losing his job, was it possible he was lying about Mr. Zhu being dead when he went into the lab? Maybe he killed him. I was certainly going to make that case to Chief Inspector Ji. Maybe, just maybe, John would be free on bail. The lab sure must have been a busy place that afternoon with all the people coming and going, and yet, no one saw anything! An early morning call to Chief Inspector Ji went on my to-do list.

Once again, sleep eluded me as I tossed and turned for the rest of the night. Things were beginning to move very fast. Now, if Gary and the Salisbury police could locate the guy the Hacketts identified and get a confession, as well who was behind the rigging of Gregory's ultralight, we might have found the killer of Mr. Zhu.

Chapter 24

My sleep was interrupted the first time at 1:00in the morning on Wednesday, June 4, when Terri called me.

"Hello, honey, did I wake you?" she asked in her sweetest voice, fearing she had.

"I was in bed but not a sleep," I said. "How did the funeral go?"

"It was very sad," Terri answered. "There was a large turnout from WMT, but not many people outside the company. The press was there, but I am not sure what their angle was? The reporter seemed more interested in digging up some dirt on the crash and Gregory than showing any sympathy. You know, in the end, he must have led a lonely life with no wife, children, or parents around him."

"From what I understand, he put everything he had into growing the business," I answered. "I can't remember ever having a personal conversation with him. It was always strictly business, but for some people, that's enough. How did Ann take it?"

"She was okay. She was very emotional at first, but did better as the morning went on."

"What are your plans for the rest of the day?" I asked.

"Ann is meeting with some clients, and I am going to crash around the pool at the hotel. I need to regenerate," Terri said.

"Wish I was there with you!"

"Me, too!"

"I do have some incredibly important news for you," I said. "Billy came to my room tonight wanting to tell me something he did not feel comfortable discussing in the plant. It seems he walked into the QC lab sometime after the murder of Mr. Zhu, found the eighth life test report, and took it with him."

Terri shouted into my ear, "You have got to be kidding! You mean to tell me he could have helped us avoid all of this and maybe have John released on bail, and he hid the report?"

"That's right, babe," I answered. "He told me he was afraid of losing his job if someone saw the report and the results were as bad as Mr. Zhu said they were, so he took it home and hid it in his apartment. He was hoping Gregory would start production because the extended life test could not be found and therefore not see any need to hold up production."

"What happens now? How are you going to get your hands on it? When will he be giving it to you and where?" Terri asked.

"Billy said he would be in Shanghai tomorrow, and he would meet me in People's Square tomorrow night at nine o'clock with the report. Your flight arrives at two in the afternoon. Why don't I go on to Shanghai tomorrow morning and get us and Ann a room at the Shangri-La, and you both can come straight to the hotel after you clear customs and have your baggage."

"Sound good to me," Terri said. "I can't wait to see you."

"Me, too, babe. Being in China without you is no fun!"

"You don't think Billy could have killed Mr. Zhu, do you?"

"I really don't think so, but I am going to do my best to convince Chief Inspector Ji we have enough evidence now for him to at least think about someone other than John being the killer," I said. "Hell, you never know, though!"

"You must be exhausted but at the same time wired," Terri said. "Honey, please try to get some sleep. You will need to be at your best tomorrow."

"I will. Give my best to everyone there. See you tomorrow."

"Before I go, is it okay to tell Dennis and Ann what has happened?" Terri asked.

"I don't see why not," I answered. "Have Dennis call me if he has any questions."

"I will do that, honey; now please, do your best to get some rest."

"I will. Call me on my cell when you land."

"Okay, talk to you later."

"Fly safe," I said as I hung up. I always wondered why people said "fly safe" to anybody other than the pilots, but oh well.

This time, my cell rang at 3:00 in the morning. "This had better be good," I said to myself as I reached for the phone.

"This is Rick," I said in the middle of a yawn.

"Wake up, my friend! It's Gary, and I have great news!"

My nerves began another ride up and then down the Eliminator roller coaster at Carowinds Theme Park near Charlotte. I was not sure how many jolts they could take.

"Gary, this had better be good," I said. "I only take good news calls after two in the morning!"

"Oh, this is better than good news; this is colossal news," Gary answered, laughing. "We got the son of a bitch around noon. We traced him through his rental car. Jim's guys picked him up at his apartment. Not a very bright guy!"

"Who is he?" I asked.

"His name is Jack Spalding," Gary said. "He is someone the Charlotte police have had an interest in for a few years. He travels in the right circles, but never seems to have a real profession. He's six five, blond, good looking, and well spoken. You know, the type that preys on unsuspecting and vulnerable women. He has also been known to associate with some local shady characters."

"Has he confessed?" I asked.

"No, not yet, but he doesn't have an alibi and we have a copy of the rental car receipt and the Hacketts are on the way over here to identify him," Gary said. "It is only a matter of time until he tells us everything about who hired him."

"That's terrific, Gary," I said. "You guys have really moved quickly on this one. I have news for you, too," I said.

I spent the next few minutes relaying my conversation with Billy and the plan to meet him tonight at People's Square to get the test report. Gary asked if finding Gregory's potential killer and the extended life test report would be enough to spring John. I told him my plan to call Chief Inspector Ji and said his news would be more ammunition to have John released on bail.

Gary said, "I will call you as soon as this asshole breaks."

"Please do!" I said. "Talk to you soon."

I hung up the phone and made my last attempt of the night to sleep. It was 10:00, later that morning, when the maid rang the doorbell and woke me. There was no need to check out of the hotel, so I packed enough clothes for a few days in Shanghai. I could just make an 11:00 morning train arriving in Shanghai at noon. The taxi ride was forty-five minutes from the train station. That gave me just enough time to register and get settled in our room before Terri would arrive from Charlotte.

"Chief Inspector Ji, please. Chief Inspector Ji, it's Rick Watson. I wanted to bring you up to date on what has happened over the last twenty-four hours."

I quickly relayed my conversation with both Billy and Gary, making my case for John's release.

"You are beginning to make headway, Mr. Watson. I suggest you contact Mr. Alworth's attorney and resubmit your bail request," Chief Inspector Ji replied.

"Are you saying you will not block a request this time?"

"All I can say is there may be enough circumstantial evidence to bring question on who the killer is," he replied.

"Thank you, Chief Inspector, I understand," I replied.

My train was on time, and I arrived at the Shangri-La as I predicted. The room, though not a suite, was comfortable and had a river side view. There was a very comfortable chair sitting directly in front of the window. I decided I needed some think time and that was the best spot in the room for doing it.

I was jolted awake by my cell ringing. I looked at the time on the face of my cell. It was 2:45 in the afternoon. I had been asleep for over an hour.

"Hello."

"Hi, honey, it's me. We are through customs and walking to the taxi waiting area. I should be at the hotel no later than four. Can't wait to see you!"

"Me, too, love you. See you soon," I replied.

"Wow," I said to myself. I must be tired to fall asleep in a chair no matter how comfortable!

Terri was right on time at 4:00when she knocked on the door of our room. I helped her in with her suitcase and spent the next several minutes hugging and kissing her.

"Boy," she said. "You really did miss me!"

"Of course I did!" I replied.

We spent the next two hours catching up on everything that happened in China and the United States. She told me the board of directors had asked Dennis to become the interim CEO until all of the events surrounding Gregory's death and the stoppage of production of the new product was cleared. She said she and Ann had congratulated him that morning right after the board meeting.

I asked her why they wanted to stay to hear the results of the board of directors meeting, and Terri said it was Ann's idea, plus they could not have made a flight to Shanghai on Monday anyway.

Terri suggested we call Ann's room and ask her to join us for an early dinner in the hotel prior to my meeting with Billy. We agreed to meet in the lobby restaurant in the western section by 6:30.

Over dinner was a good time to share the events of the past twenty-four hours with Ann, including my conversation with Chief Inspector Ji. Ann was thrilled and said she was going to call Mr. Jianguo right after dinner.

I told her about Billy coming to my room and confessing he had removed the original test report for the extended eighth life test from Mr. Zhu's office. I told her I was going to meet Billy at 9:00 in People's Square in front of the theater to pick up the original test report on the unapproved eighth life test. I explained that if the test was correct, the results would be devastating for WMT.

Ann became somewhat distracted as I was talking.

"Honey, what's wrong?" Terri asked her.

"Oh, nothing. Just wishing John was here. If you both will excuse me, I am going back to my room to call Mr. Jianguo. Maybe there is something he can do tonight to get John released."

Terri finished dinner and returned to our room at 7:20. I didn't notice it, but I was pacing back and forth across the room and watching the clock impatiently, waiting for the time to leave for People's Square, when my cell phone rang. It was 7:30.

"Hello," I said.

"I hope you are sitting down, my friend!" Gary said. "Mr. Spalding decided to save, at least, some of his skin this morning. He confessed to rigging Gregory's ultralight and gave us the name of who hired him. It seems at one time he was going to marry this person but it didn't work out. He has always had the hots for her. She approached him, turned him on, and lead him to believe if he would do this for her, she would be his again."

"Come on, Gary, I'm not in the mood. It's been a hard day," I said.

"Okay, okay," Gary said. His voice was that of a kid whose surprise birthday party was no longer a surprise. "The person who hired him and also probably killed your Mr. Zhu is none other than Ann Alworth!"

"*Oh my God!*" I shouted, startling Terri. "Are you sure?"

"Absolutely!" Gary replied. "It seems good ole Mr. Spalding has always had the hots for Mrs. Alworth long before she met John. Evidently, she promised Spalding a half a mil. if he would get rid of Gregory."

"Jesus, that's a lot of money! Did he say why she offered him so much?"

"He said something about her having a lot of stock options in WMT and that she would be rich very soon," Gary replied.

"What the hell! . . . *Oh shit!* Terri, where is Ann? Call her room, now!"

"Hold on, Gary."

Terri dialed Ann's room. "There's no answer."

"Try her cell!"

"No answer there, either," Terri said.

"Gary, got to go. Call you back later," I shouted into the phone.

"Ann's on the way to People's Square," I said. "She is going to kill Billy and get the test report!"

"That can't be," Terri said. "There must be some mistake."

"No mistake, honey," I said, heading for the door. "If Ann had stock options large enough to pay Spalding five hundred thousand dollars to kill Gregory, she has a better reason to kill Billy and destroy the report!"

"I'll keep calling her cell," Terri said. "Maybe she will answer."

We reached the front of the hotel and hailed a taxi before the doorman could react. I shouted at the doorman and told him to tell the driver to take us to the theater in People's Square as fast as possible. I told him there would be a big tip if we had a quick trip.

The driver gave a big smile and gunned the taxi away from the hotel.

I called Shanghai police headquarters and asked to be connected to the chief superintendent of police, Feng Shou. The desk sergeant said he was off duty and for me to call back in the morning. I politely explained who I was and that the call was an extreme emergency and I desperately needed to talk to the chief superintendent.

What I said must have worked, because the next voice I heard was Chief Superintendent Shou's.

"Chief Superintendent Shou, this is Rick Watson. I sure hope you remember me."

"I remember you, Mr. Watson," he said in a haughty voice. Obviously, Chief Superintendent Shou hadn't mellowed over the past month and a half. "What is the emergency you are having?"

I quickly summarized the events of the past month and a half with detail around the last twenty-four hours, including my belief that Ann Alworth was on the way to People's Square to commit murder for the second time.

"I will notify all units and officers in that area and am on my way to meet you there," he said.

I looked at my watch. It was twenty minutes until 9:00. Our driver was making excellent time, and I knew we would arrive at the theater in ten minutes.

Terri kept calling Ann's cell with no response.

I asked Terri to give the driver fifty dollars American money and be prepared to jump out of the taxi as soon as the tires stopped rolling.

People's Square is a very popular place in Shanghai for residents and tourists alike. Tonight, the theater was dark, so the crowds were smaller than usual.

Neither Ann nor Billy were anywhere in sight when I completed my first scan of the area. Maybe we were wrong about Ann or here early before Billy had arrived, or maybe he changed his mind. I looked at my watch. It said 9:05.

"Rick, look over there!" Terri said, pointing to a dimly lit pathway leading into a small wooded area. "There they are!"

Ann was walking behind a very frightened Billy. She appeared to be nudging him with something she had in her hand. She was continually looking in all directions, making sure no one was paying attention to them.

I began running toward them and yelled as I got close enough for Ann to hear me. "Ann, stop. It's all over. We know all about you hiring Jack Spalding to kill Gregory!"

When Ann heard the name Jack Spalding, she turned, and for the first time, I saw the gun she had been pointing at Billy. Before I could react, she fired!

It was like a movie scene in slow motion. I saw the bullet coming toward me but could not move fast enough. The bullet hit me on the side of the head, sending me sprawling across the concrete pathway into a tree. The bullet lodged into a second tree right behind me. I faintly heard Terri scream and begin to run toward me.

"Damn, that hurt," I said to myself, rejoicing I was still alive. I was able to sit up and check out the wound. Thank God Ann was a bad shot. The bullet only grazed the side of my head.

"Oh my God," Terri said with panic in her voice. "Honey, all you all right? God, please let him be all right!" she pleaded.

"I'm okay, I'm okay," I said.

Before Terri could respond, Ann raised her gun to fire the second time when someone shouted a warning in Chinese. In a panic, Ann turned and pointed the gun at the voice. She looked as if she was prepared to shoot. A Shanghai police constable walking his normal beat appeared under a dim street light. He yelled the second time, and when Ann failed to respond, he fired. Ann fell to the ground in a heap.

Billy didn't know what to do, so he ran around in circles like a circus clown until he calmed enough to run to me.

"Mr. Rick, you okay?" Billy asked.

"I'm okay. The bullet only grazed the side of my head. Terri, go check on Ann."

Chief Superintendent Shou arrived on scene with sirens blazing just as Terri and the constable reached Ann.

"She's still alive!" Terri said. "Please, get an ambulance!"

The constable understood enough English to understand what Terri was saying. He radioed somewhere and requested an emergency vehicle.

Chief Superintendent Shou quickly took command of the scene upon his arrival.

"Mr. Watson, are you all right?" he asked, looking at the wound on the side of my head.

"I'm okay. Take care of her first, please."

Chief Superintendent Shou nodded and reconfirmed with the constable an emergency vehicle was on its way.

Within minutes, the emergency vehicle arrived. The paramedics stabilized Ann and loaded her into their vehicle for a quick trip to Yan An Hospital in central Shanghai.

"Mr. Shou, please, ask them how she is?" I asked.

He spent a minute in conference with the paramedics and said, "Not too good. They must hurry. You please go with them?" he said, pointing to the ambulance.

"No, I'm all right," I said. "Will you please drive us to the hospital?"

"Yes, please, this way."

I had forgotten about Billy. He was standing off to the side of the street sobbing. "Billy, it's okay. I'm fine, and Ann will be okay."

"Mr. Rick, I so sorry. It all my fault!"

"Forget it, Billy. You have done the right thing. We now have the report and everything will be worked out. Please, go home. We will talk tomorrow."

Billy nodded and handed me the report before turning and walking away, saying goodbye in a very small voice.

There was no time to worry about Billy now. I was confident he would make his way home. I would work with him later to ensure he didn't try to do anything stupid over what had just happened.

Chief Superintendent Shou's driver was expertly maneuvering through traffic as he reached a main road that led to the hospital.

My cell phone had survived my fall and was still in my pocket. I called Chief Inspector Ji at the Changzhou police headquarters. The desk sergeant said Chief Inspector Ji was not on duty but could be reached in case of emergency. I explained who I was and why I needed to talk to him urgently. The desk sergeant asked me to hold. He returned in less than a minute.

"Please, sir, your number," he said.

I gave him my cell number and hung up as he requested.

My cell rang almost instantly.

"Mr. Watson, my desk sergeant tells me you urgently need to speak to me," he said.

"Chief Inspector Ji, I don't have much time, but I need you to arrange for John Alworth to be brought to the Shanghai Yan An Hospital. Ann, his wife, has been shot. I am with Chief Superintendent Feng Shouof the Shanghai police on the way to the hospital now. Ann Alworth is responsible for the deaths of Mr. Zhu and Gregory Brightson. I will give you all the details when you arrive. Please, hurry! Ann is in critical condition."

"I believe I can arrange for a police helicopter," Chief Inspector Ji said. "I will call you back very soon."

We arrived at the hospital right behind the ambulance. Chief Superintendent Shou insisted I have someone look at my head wound while he and Terri ensured Ann was taken to an emergency room.

My respect for Chief Superintendent Shou grew as everyone in the emergency room began to scurry around at his commands. A young doctor appeared and escorted me to an examining room. He examined, treated, and bandaged my wound within minutes. Another medical team wheeled Ann into the emergency room next to mine.

Terri and Chief Superintendent Shou were standing near the bed behind the medical team working on Ann. A doctor extricated himself from the rest of the team and spoke to Chief Superintendent Shou in Chinese.

"Mr. and Mrs. Watson, there is nothing we can do here," Chief Superintendent Shou said. "The doctor will advise us as soon as possible. He suggests we wait in the family waiting room down the hall."

My cell rang as we were walking down the hall.

"Mr. Watson, this is Chief Inspector Ji. I have secured a helicopter, and we shall be arriving at the hospital within the hour."

"Thank you, Chief Inspector, for believing me and acting so quickly," I said.

"This time, it is my pleasure, Mr. Watson. Mr. Alworth and I will see you soon."

"Have you told Mr. Alworth what has happened?" I asked.

"Yes, I have. Unfortunately, he is in a state of shock at this moment and hasn't said much."

"Thank you again for believing me," I said, hanging up my cell.

Watching the clock move was madding as Terri and I continually paced up and down the hall. What was taking so long? Was Ann going to live? Somebody had to tell us something soon! It had been over an hour since we arrived at the hospital.

A commotion began outside the elevator door. We all looked and recognized Chief Inspector Ji and John Alworth hurrying down the hall toward us.

"My God, Rick, are you all right?" John asked as he approached me, looking at the bandage around my head. "And where is Ann? Is she going to be all right?"

"I'm okay, but we are not sure about Ann. She tried to kill me and would have probably attempted to kill a Shanghai constable if he had not shot her first," I said.

"My Ann," John screeched as her looked at me in disbelief. "My Ann killed someone?"

"Here," I said, pointing to a chair at the end of the hall. "Sit down and I'll explain all I know about what happened this evening and the past week."

John sat with his head hung between his legs as Terri and I did our best to explain Billy confessing he had the original extended eighth life test report, Jack Spalding's confession implicating Ann in the death of Gregory Brightson, Ann's attempt to kill Billy in People's Square, and my belief that she was responsible for killing Mr. Zhu even though I had no physical proof of that one at this moment.

I wasn't sure John was receiving and comprehending what we were saying. He displayed no physical movement while we were speaking, but finally raised his head with a shocked, questioning look.

"But all of this makes no sense!" John said. "What motive would she have, for God's sake, to do something like this?"

"Jack Spalding, during his confession, said Ann offered him five hundred thousand dollars to kill Gregory. She told him she had big WMT stock options worth millions as soon as the company went public."

"How can that be, Rick? She didn't work for the company, and I certainly didn't have that many options myself," John asked.

"That's all I know right now. Maybe we will learn more when we talk to Ann."

One of the emergency room doctors approached us as I was finishing talking to John.

"We have stabilized the patient and moved her to an ICU room on the sixth floor," the doctor said.

"When will we be allowed to talk to her?" Chief Superintendent Shou asked.

"Within the hour, assuming her condition is steady, but only for a few minutes at a time and not a room full of people, understand?"

"Yes, Doctor," Chief Superintendent Shou said.

We made our way to the sixth floor waiting room. Both police officers separated themselves from us and began a discussion of their own in Chinese. I looked at my watch. It was a few minutes past midnight. I decided this was a good time to call Dennis Dureno to bring him up to date on the events of the evening while Terri was doing her best to comfort John, who sat weeping on a long couch positioned under a large window.

Dennis answered after the third ring. "Dureno," he said.

"Dennis, this is Rick Watson," I said.

"Rick, it's after midnight in China," he said. "What's up?"

My efficiency increased every time I relayed the events of the evening and past twenty-four hours. This time was no exception.

"Holy shit, Rick. Now, some of what I have been finding in Gregory's office and conversations with Ann make sense," he said.

"What things?" I asked.

"I found documents in Gregory's safe where he had secretly given Ann Alworth twenty thousand share options of WMT," he said. "And I found e-mails between him and Ann on his private e-mail account, as well as receipts in his desk of hotel bills from trips they took together. It's pretty clear something was going on between them that was not business."

"That certainly clears up most of the questions I had about some of the loose ends here," I said.

"She also insisted I meet with her on Monday after the funeral and before the board of directors meeting," Dennis continued. "She was pushing me, once I was named CEO, to start production of the new product in China and keep on schedule for going public. I found it odd, but attributed it to her zeal to help John."

"I can guess the rest," I said. "We know she met John in the QC lab the day of the murder. John told her about the extended eighth life test report Zhu had run and the consequences. She must have somehow removed John's gun from his briefcase, returned later after everyone was gone except Zhu, and killed him in order to get her hands on the report and destroy it before it became public."

A doctor interrupted my conversation. "You may see the patient now, but only for a short time."

"Dennis, I have to go; the doctor says we can see Ann. I will call you back later."

After a short discussion, it was decided only Chief Inspector Ji and John would see Ann. It was important for John's sake someone from either the Shanghai police or the Changzhou Police was there to ask questions.

Terri, Chief Superintendent Shou, and I waited outside the room in the hall. Fifteen minutes had past when several alarms rang out from inside the room. We quickly stepped aside as the doctor and his medical team rushed into the room. Chief Inspector Ji immediately rushed out of the room.

"Not good!" he said with a sad look on his face.

It was only a few minutes until John walked out of the room with tears flowing down both of his cheeks. Terri grabbed him and held him to her in an unyielding hug. John held on, continuing to cry. Finally, he raised his head and said, "She's gone! My Ann's gone!"

Somehow, we shuffled back down the hall to the waiting room to allow John time to compose himself in private.

Terri and I held each other, really not believing what had happened over the past twenty-four hours. We both knew our job now was to support our best friend by getting him home and away from all of this.

John said, "I want you all to know I loved her, no matter what she did!"

Chapter 25

It was Thursday evening, June 5, the next time Terri and I talked to John. He called and wanted us to meet him in his room at the Holiday Inn. We all remained in Shanghai to await notification from the Changzhou police that all charges against John had been dropped and that we had the necessary paperwork giving John permission to transport Ann's body back to the United States. Dennis Dureno had worked diligently to ensure all legal issues and resultant costs were covered by WMT.

When we reached his room down the hall from ours, John told us he had received word all charges had been dropped by the Changzhou police department as well as a personal apology from Chief Inspector Ji.

The release of the body was being handled by Lu, Wxi, and Hoa, and John said he expected everything to be in place by tomorrow. He was planning to make flight arrangements back to the United States once he received the paperwork.

"Guys, I don't know how I will ever be able to thank you for what you have done for me. Without your help, I would probably have been found guilty of Zhu's murder, and Ann would have gotten off scot-free."

Terri gave him a hug and said, "Johnny, you know how much we love you! Rick and I are just glad it's all over, right, honey?"

"Absolutely," I replied. "We spent a lot of money while we have been here, and I need him home to win most of it back!"

John laughed at me and said, "You wish!"

"John, may I ask what happened in Ann's room before she died?" Terri said.

"She admitted to having an affair with Gregory and to taking my gun from my briefcase and using it to kill Zhu. She said she believed she had convinced Dennis to restart production after he was named CEO of the

299

company and needed to kill Billy to get the report. She said she could not find it in the QC lab after she killed Zhu. I never thought my wife could have been so devious to pull something like that off! I guess you never know, do you!"

"I almost forgot," I said. "I have the report on the eighth life test that Billy gave me. I wanted you to take a look to see what Zhu was talking about and to determine if he was right about the product creating fatalities." I handed the report to John, who immediately began scouring through each page of data in an attempt to see what Zhu had seen.

It didn't take long before John looked up from the report with a look of disbelief and disgust on his face. "Son of a bitch! You're not going to believe this. Early on in the building of the first seven production systems, receiving inspection detected a shipment of defective castings from our casting supplier. The supplier had mistakenly plated a small shipment of castings with cadmium."

John continued, "The castings were rejected and sent to the reject area to await an MRB decision to send them back to the supplier. It appears when Mr. Zhu decided to run an eighth life test, there were no acceptable castings in the plant. Someone must have placed rejected castings in the kit of material, not realizing what could happen during the life test. It is apparent no one recognized rejected castings being installed in several places the system. Under high temperature, cad plating reacts with treated water and, under the right circumstances, could cause illness to those who drink the contaminated water."

"Was Zhu right about the possibility of someone dying by drinking the potentially contaminated water?" I asked.

"No, drinking contaminated water caused by cad plating has never been known to be fatal in any case on record that I am aware of," John replied. "Zhu was right, however, that during a normal life test, this condition would not manifest itself; however, an extended period under high heat is where the problem might occur. After reading this report, it is obvious that's what happened here, leading Zhu to make his erroneous conclusion."

"John, why do you think Zhu decided to run the eight life test? All seven previous tests were normal and met specification," I asked.

John answered, "I guess that is something we will never know. I agree with you the first seven tests were all to spec. I guess it will remain a mystery why he decided to run an eighth life test and to extend the test time. Maybe he saw something we do not see in the data in one of the first seven life tests and decided to run an eighth life test and extend it."

"You mean to tell me everything that has happened could have been avoided if Zhu had only talked to you about his suspicions when he first saw what he thought he saw or had given you the eighth life test report in the beginning?" I said in utter disbelief.

"His fear or stupidity led to wasting the lives of three good people, including his own, and almost destroyed a company!" I found myself yelling at nobody.

"I'm afraid so, and I lost the love of my life," John said as he began to sob uncontrollably.

"Jesus Christ," I said. I fell back on the couch at the end of the room in complete disbelief. Terri joined me almost in a state of shock.

"All of this for nothing!" she said.

EPILOGUE

It was three weeks later on Wednesday, June 25, at 9:30 in the morning when John called. Terri and I had completed our morning run along one of our favorite paths along the lake and were enjoying a glass of sweet tea on the screened-in porch. Vito was in his normal position on someone's lap, and Annabell was chewing on her favorite ball.

The house telephone rang. Terri answered it and put it on speaker. "It's John."

"Hey, guys, I wanted to update you on what is happening at WMT. We are back on schedule to release the product a little over a month past our original schedule. Every retest is complete, and the product passed with flying colors. The board, this morning, named Dennis Dureno the permanent CEO, and I am now the vice president of operations. How about that!"

"Well, congratulations!" Terri and I shouted simultaneously. "It couldn't happen to better guys!"

"Thanks," John replied. "We have a lot of work ahead of us. I guess the next time I will see both of you is next week when my team kicks your team's ass at the charity golf tournament!" John said, laughing.

"Not this time, my friend," I replied. "I assure you, Clemson and my team will do the ass kicking!"

"Get your money ready, sucker. I want it all!" John said and hung up.

The Clemson/Carolina Charity Golf Tournament played every July is one of several hotly contested athletic events staged yearly between the heated rivals. This year's tournament was scheduled for the weekend of July 25. It was the hottest time of the year in the Charleston area, but the participants used the heat as a badge of courage!

Normal attendance ranged between eight hundred to one thousand participants. This year's match was held at Wyboo Golf Course, located on Wyboo Plantation on Lake Marion. I hoped I would have an advantage over John this year because Wyboo was my home club.

The tournament was hard fought as usual, with the normal amount of trash talk. Shouts could be heard across the course when one team or the other gained an advantage. John and I contested each hole against each other as if it was the last hole of the Masters.

Like last year, the tournament championship and the largest charity donation boiled down to the eighteenth hole with my Clemson team leading by one stroke. Personally, I was one up on John and not letting him forget it as he teed off. Clemson would win this year, period!

John and I continued to trash talk before and after each shot leading to the eighteenth green, a long par-four dog leg right. John hit a tremendous second shot, leaving a short chip to the green and positioning him for a birdie putt.

My third shot landed three feet from the cup, leaving me a short putt for a par. I felt relieved that a par putt would mean my team would at least be in a position to win the tournament championship or to compete with John's team in an extra hole playoff.

Winning the tournament would give Clemson Charities the largest portion of the fees and sponsorships; however, bragging rights for the next twelve months, plus I would win my side bet with John—*priceless*!

Walking up the eighteenth fairway was our Masters. We could see a very huge crowd had gathered surrounding the eighteenth green in anticipation of our putts. Few participants had left the course.

Like last year, everyone deferred to us by completing their putts, ensuring that what John and I did would decide the championship.

John was away, which meant he had to putt first. I was waiting for John to putt when he said, "Go on, loser, go ahead and putt out so I can watch you bawl like a Clemson baby when I sink this birdie putt. It's only thirty-five feet. A piece of cake!"

"A cake you will never enjoy," I replied. My putt was only three feet with no break. Making it was no problem. I reached into the cup to retrieve my par putt to the cheers of the Clemson crowd. "I can't wait until that loooong putt of yours rolls off the green when you choke."

I walked behind John to review his putt. I had made that putt myself a few times in other rounds at Wyboo and knew there was a substantial break left at the cup after the ball climbed the hill between the cup and John.

John lined up his putt and said, "Read it and weep, sucker!" His stroke was firm and his ball climbed the hill with enough speed to head toward the cup. It fell over the crest and gained speed as it rolled downward toward the cup.

Almost a thousand people surrounding the green held their collective breath. Clemson participants prayed for John to miss, and the Carolina team was ready to breakout a huge cheer if the ball went into the cup!

"Come on, come on, miss the break," I said hopefully to myself as John's ball slowed as it rolled toward the hole. The break was on the left side of the cup, and John had read the break or at least partially. His ball was rolling toward the cup from the left side, but was it out far enough?

"No break, no break!" I yelled. The yell escaped as if it had a mind of its own.

John began to laugh. "What's the matter, boy? You worried!"

Maybe I was in luck; the ball was really slowing now. Maybe it wouldn't even make it to the cup, I thought. *No, it was almost there. It was turning and moving slower now! The ball turned left toward the cup! Oh no! Oh yes! The ball reached the lip of the cup; the crowd around the green went wild . . .!*